The Palgrave Macmillan Animal Ethics Series

Series editor

In recent years, there has been a growing interest in the ethics of our treatment of animals. Philosophers have led the way, and now a range of other scholars have followed from historians to social scientists. From being a marginal issue, animals have become an emerging issue in ethics and in multidisciplinary inquiry. This series explores the challenges that Animal Ethics poses, both conceptually and practically, to traditional understandings of human–animal relations.

Specifically, the Series will:

- provide a range of key introductory and advanced texts that map out ethical positions on animals;
- publish pioneering work written by new, as well as accomplished, scholars; and
- produce texts from a variety of disciplines that are multidisciplinary in character or have multidisciplinary relevance.

Titles include:

AN INTRODUCTION TO ANIMALS AND POLITICAL THEORY
Alasdair Cochrane

Forthcoming titles

ANIMALS, SUFFERING AND PHILOSOPHY
Elisa Aaltola

HUMANS AND ANIMALS: THE NEW PUBLIC HEALTH PARADIGM
Aysha Akhtar

HUMAN ANIMAL RELATIONS: THE OBLIGATION TO CARE
Mark Bernstein

ANIMAL ABUSE AND HUMAN AGGRESSION: THE SAME OR DIFFERENT PHENOMENA?
Eleonora Gullone

ANIMALS IN THE CLASSICAL WORLD: ETHICAL PERCEPTIONS
Alastair Harden

ANIMAL SUFFERING AND THE PROBLEM OF EVIL
Nicola Hoggard Creegan

POWER, KNOWLEDGE, ANIMALS
Lisa Johnson

ANIMAL EXPERIMENTS: EVIDENCE AND ETHICS
Andrew Knight

CRITICAL ANIMAL STUDIES: AN INTRODUCTION
Dawne McCance

D1627704

POPULAR MEDIA AND ANIMAL ETHICS
Claire Molloy

ANIMALS, EQUALITY AND DEMOCRACY
Siobhan O'Sullivan

SOCIAL WORK AND ANIMALS: A MORAL INTRODUCTION
Thomas Ryan

AN INTRODUCTION TO ANIMALS AND THE LAW
Joan Schaffner

UNIVERSAL ANIMAL RIGHTS: WINNING THE ETHICAL DEBATE
David Sztybel

The Palgrave Macmillan Animal Ethics Series
Series Standing Order ISBN 978–0–230–57686–5 Hardback
978–0–230–57687–2 Paperback
(*outside North America only*)

You can receive future titles in this series as they are published by placing a standing order. Please contact your bookseller or, in case of difficulty, write to us at the address below with your name and address, the title of the series and the ISBN quoted above.

Customer Services Department, Macmillan Distribution Ltd, Houndmills, Basingstoke, Hampshire RG21 6XS, England

An Introduction to Animals and Political Theory

Alasdair Cochrane
London School of Economics, UK

© Alasdair Cochrane 2010

All rights reserved. No reproduction, copy or transmission of this
publication may be made without written permission.

No portion of this publication may be reproduced, copied or transmitted
save with written permission or in accordance with the provisions of the
Copyright, Designs and Patents Act 1988, or under the terms of any licence
permitting limited copying issued by the Copyright Licensing Agency,
Saffron House, 6-10 Kirby Street, London EC1N 8TS.

Any person who does any unauthorized act in relation to this publication
may be liable to criminal prosecution and civil claims for damages.

The author has asserted his right to be identified
as the author of this work in accordance with the Copyright,
Designs and Patents Act 1988.

First published 2010 by
PALGRAVE MACMILLAN

Palgrave Macmillan in the UK is an imprint of Macmillan Publishers Limited,
registered in England, company number 785998, of Houndmills, Basingstoke,
Hampshire RG21 6XS.

Palgrave Macmillan in the US is a division of St Martin's Press LLC,
175 Fifth Avenue, New York, NY 10010.

Palgrave Macmillan is the global academic imprint of the above companies
and has companies and representatives throughout the world.

Palgrave® and Macmillan® are registered trademarks in the United States,
the United Kingdom, Europe and other countries.

ISBN: 978–0–230–23925–8 hardback
ISBN: 978–0–230–23926–5 paperback

This book is printed on paper suitable for recycling and made from fully
managed and sustained forest sources. Logging, pulping and manufacturing
processes are expected to conform to the environmental regulations of the
country of origin.

A catalogue record for this book is available from the British Library.

Library of Congress Cataloging-in-Publication Data
Cochrane, Alasdair, 1978–
　An introduction to animals and political theory / Alasdair Cochrane.
　　p.　cm. — (The Palgrave Macmillan animal ethics series)
　Includes bibliographical references and index.
　ISBN 978–0–230–23925–8 (hbk), 978–0–230–23926–5 (pbk)
　1. Animal welfare.　2. Human–animal relationships.　I. Title.
　HV4708.C63 2011
　179′.3—dc22
　　　　　　　　　　　　　　　　　　　　　　　　　　　　2010027496

10　9　8　7　6　5　4　3　2　1
19　18　17　16　15　14　13　12　11　10

Printed and bound in Great Britain by
CPI Antony Rowe, Chippenham and Eastbourne

Contents

Series Preface   vi

Acknowledgements   viii

1 Introduction: Animals and Political Theory   1

2 Animals in the History of Political Thought   10

3 Utilitarianism and Animals   29

4 Liberalism and Animals   50

5 Communitarianism and Animals   72

6 Marxism and Animals   93

7 Feminism and Animals   115

8 Conclusions   136

Notes   147

Bibliography   157

Index   163

Bedfordshire County Council	
9 39280536	
Askews	
179.3 COC	

Series Preface

This is a new book series for a new field of enquiry: Animal Ethics.

In recent years, there has been a growing interest in the ethics of our treatment of animals. Philosophers have led the way, and now a range of other scholars have followed from historians to social scientists. From being a marginal issue, animals have become an emerging issue in ethics and in multidisciplinary enquiry.

In addition, a rethink of the status of animals has been fuelled by a range of scientific investigations which have revealed the complexity of animal sentiency, cognition and awareness. The ethical implications of this new knowledge have yet to be properly evaluated, but it is becoming clear that the old view that animals are mere things, tools, machines or commodities cannot be sustained ethically.

But it is not only philosophy and science that are putting animals on the agenda. Increasingly, in Europe and the USA, animals are becoming a political issue as political parties vie for the 'green' and 'animal' vote. In turn, political scientists are beginning to look again at the history of political thought in relation to animals, and historians are beginning to revisit the political history of animal protection.

As animals grow as an issue of importance, so there have been more collaborative academic ventures leading to conference volumes, special journal issues, indeed new academic animal journals as well. Moreover, we have witnessed the growth of academic courses, as well as university posts, in Animal Ethics, Animal Welfare, Animal Rights, Animal Law, Animals and Philosophy, Human-Animal Studies, Critical Animal Studies, Animals and Society, Animals in Literature, Animals and Religion – tangible signs that a new academic discipline is emerging.

'Animal Ethics' is the new term for the academic exploration of the moral status of the non-human – an exploration that explicitly involves a focus on what we owe animals morally, and which also helps us to understand the influences – social, legal, cultural, religious and political – that legitimate animal abuse. This series explores the challenges that Animal Ethics poses, both conceptually and practically, to traditional understandings of human–animal relations.

The series is needed for three reasons: (1) to provide the texts that will service the new university courses on animals; (2) to support the increasing number of students studying and academics researching in animal related fields; and (3) because there is currently no book series that is a focus for multidisciplinary research in the field.

Specifically, the series will

- provide a range of key introductory and advanced texts that map out ethical positions on animals;
- publish pioneering work written by new, as well as accomplished, scholars; and
- produce texts from a variety of disciplines that are multidisciplinary in character or have multidisciplinary relevance.

The new Palgrave Macmillan Series on Animal Ethics is the result of a unique partnership between Palgrave Macmillan and the Ferrater Mora Oxford Centre for Animal Ethics. The series is an integral part of the mission of the Centre to put animals on the intellectual agenda by facilitating academic research and publication. The series is also a natural complement to one of the Centre's other major projects, the *Journal of Animal Ethics*. The Centre is an independent 'think tank' for the advancement of progressive thought about animals, and is the first Centre of its kind in the world. It aims to demonstrate rigorous intellectual enquiry and the highest standards of scholarship. It strives to be a world-class centre of academic excellence in its field.

We invite academics to visit the Centre's website www.oxfordanimal ethics.com and to contact us with new book proposals for the series.

Andrew Linzey and Priscilla N. Cohn
General Editors

Acknowledgements

I would like to thank all those who have helped to see this book into publication. First of all, I would like to offer my thanks to Andrew Linzey and Priscilla Cohn for establishing this exciting series. The editorial team at Palgrave deserve high praise for handling the transformation of the work into a finished book so smoothly: special thanks to Priyanka Gibbons, Melanie Blair and David Joseph. I must acknowledge the enormous debt I have to all those teachers who have helped shape my understanding of political theory, including: Geraint Williams, Anthony Arblaster, James Meadowcroft, Mike Kenny, Janet Coleman, Paul Kelly and Cecile Fabre. I must also thank Camillia Kong for introducing me to the work of Frances Hutcheson. I have discussed many of the ideas in this book and its structure with Robert Garner, whose important research on animals and political theory has been invaluable. Finally, thanks to the friends and family who have motivated me throughout.

1
Introduction: Animals and Political Theory

This book addresses the question of how political communities ought to govern their relations with non-human animals. The question is important because, quite clearly, all political communities must take *some* position over how animals in their society ought to be treated.[1] After all, each and every political community has important social interactions with animals, whether those animals are companions, modes of transport, sources of awe, labourers, gods, food sources or simply part of a shared environment. Political communities have to make important decisions about whether those interactions should be regulated, how, for what reasons and for what ends. And, of course, all political communities *do* make such decisions. The past few years, for example, have seen a number of interesting developments. In 2007, the Animal Welfare Act came into force in England and Wales. Often referred to as a 'Bill of Rights for Animals', the Act introduced tougher fines for animal cruelty and granted greater powers of intervention to the RSPCA. Crucially, the Act also altered the legal obligations of owners of animals in a significant way: owners in England and Wales no longer simply have to refrain from acting cruelly towards their animals, but are also legally obliged to take positive steps to ensure that their animals are well housed, free from pain and able to express normal behaviour.[2] A year later, in 2008, the Spanish parliament approved resolutions recommending the enactment of legislation which would assign to great apes the right not to be tortured, the right to life and the right to freedom. Some commentators dubbed this step as essentially entailing a parliamentary recommendation of the award of *human rights* to such animals.[3] Another year on, in the summer of 2009, Chinese lawyers began taking the initial tentative steps in drawing up China's very first piece of animal welfare legislation. Currently animals that are sold as

1

pets or for meat in China can be beaten, tortured and killed without penalty.[4]

These examples illustrate well the different stances that political communities take in terms of their treatment of non-human animals. Of course, it is possible to think from these examples that while there are differences in the ways that political communities treat animals, a similar overall trend is evident: a trend of states offering greater and better protection to animals. But while it is certainly possible to argue that animal welfare legislation is generally getting tougher, that does not necessarily correlate with a better situation for animals in reality. Take, for example, the numbers of animals raised, killed and eaten in modern societies. The global per capita consumption of meat has more than doubled between 1961 and 2007, and is expected to double again by 2050.[5] Political communities may well be instituting stronger pieces of animal welfare legislation, but they are also raising, killing and eating animals like never before. Asking how political communities ought to govern their relations with animals involves asking more than simply what laws they should enact.

This book's primary concern is not to explain *why* political communities take the stances regarding the treatment of animals that they do. Nor is it to *compare* the different types of steps that political communities have taken. Rather, the purpose of this book is to introduce and analyse the type of policies that political theorists have claimed that communities *should* take. In other words, the book addresses the 'normative' question: the question of how political communities *ought to* govern their relations with animals. For example, is the Animal Welfare Act of England and Wales a significant and positive step forward in animal welfare legislation? Or is it a desperately conservative move, entrenching the idea that it is permissible for humans to own and exploit animals for their own ends? Should Spain extend human rights to the great apes? If so, should they in fact move beyond the great apes, and extend such rights to all sentient animals? Should China introduce animal welfare legislation which is similar to that which exists in the UK, Spain and other European states? Or, should China frame quite different policies based on its own communal attitudes to animals? These types of normative questions are the concern of this book.

But it is important to note that this book is not concerned with each and every type of normative question relating to animals. After all, this book is about political theory, and political theorists are primarily concerned with the actions of political communities, rather than individuals. As such, a political theorist will be less interested in whether we each as individuals have a moral obligation to be vegetarian, a question which might, for instance, concern some moral philosophers. Instead, the political theorist will be much more interested in the question of how the state relates to the meat industry: whether it should support it, regulate it or close it down. That is not to say, however, that individual morality is of no interest to the political theorist. For one, communities are obviously made up of individuals, and the policies of states are crucially dependent upon the norms and actions of individuals. Moreover, and as we will see, some political theorists believe that the good life of individuals and the flourishing of political communities are inextricably linked. In spite of these important points, it is nevertheless useful to remember that political theory's primary concern is with how political communities should govern themselves. In other words, put in the parlance of political thinkers themselves, political theory's main concern is with 'justice'.

One of the important controversies that this book addresses is whether justice can or should include animals. As we will see, different thinkers have different views on this. However, it is important at this stage to have some idea of what including animals within justice actually involves. I want to argue that it involves two things. First of all, extending justice to animals involves recognising that the treatment of animals is a matter for political communities to *enforce*. That is to say, including animals within justice means that how we relate to animals is not merely a private or individual matter, but rather a public and political matter that states should regulate. So, for example, we can coherently say that China does not currently include pets of farm animals within justice, because it has no enforceable state legislation on how those animals ought to be treated. Secondly, however, extending justice to animals also involves recognising that the treatment of animals is something that political communities ought to enforce *for the sake of animals themselves*. After all, including animals within justice involves an implicit recognition that animals *merit* justice. For example, if a state were to enforce animal welfare legislation not out of any concern for animals themselves, but merely because the state considered it good to be compassionate, it would be wrong to say that the animals of that community had been extended *justice*. Rather, it would seem more appropriate to say that they had been extended charity or compassion. When an individual is included in *justice*, on the other hand, then that individual merits certain treatment for their own sake, and not simply for the sake of others. For example, the European Union's (EU) Treaty of Amsterdam of 1997 includes a legally binding protocol which declares that the welfare of animals ought to be considered when formulating legislation in the areas of research, transport, agriculture and the internal market. Importantly, the protocol explicitly declares that this provision is owed to animals on the basis of their sentience. As such, this protocol can be viewed as an explicit recognition by the EU that animals merit justice for their own sakes. Given the EU's and China's contrasting stances, we can see that the question of whether animals are owed justice or not is an important and contested issue. Moreover, the issue is not just contested amongst states, but also amongst political theorists.

So, what exactly do political theorists say about the animal issue? What do they say about whether animals are owed justice, and what justice for animals might actually entail? This book answers these questions by systematically discussing the most prominent schools in contemporary Western political theory and examining their implications for animals. Utilitarianism, liberalism, communitarianism, Marxism and feminism are each introduced in turn, and their contributions to the debate over animal justice are examined and evaluated. But an important note of caution is necessary here. For it is important to understand that the animal issue has been somewhat neglected in political theory. Quite simply, most political theorists have said, and continue to say, almost nothing on what political communities owe to animals. Indeed, when one looks back over the history of political thought, this silence is particularly striking. Of course, this is not to say that animals are *never* mentioned in the great works of political theory. For in actual fact, political thinkers have frequently employed animals in their work; but they have employed them as units of comparison to illustrate the exalted status of human beings. Theorists have commonly discussed animals in order to identify some unique human characteristic, and have then gone on to use that characteristic as a basis with which to build their accounts of what a just political community would look like. For example, Aristotle claimed that the unique human characteristic was the capacity for the moral and intellectual virtues. As such, he argued that the ultimate goal of political communities was to promote those virtues. For theorists like Cicero and Aquinas, on the other hand, the unique human characteristic was reason. These theorists claimed that reason gives us access to the natural law, which should in turn form the basis of the laws of the state. While for Marx, the defining human characteristic was productive labour. Moreover, he argued the full realisation of that power can only be achieved in a communist society.

In all of these examples, political theorists used animals to help build their accounts of how political communities should govern themselves.

However, none of them actually included the treatment of animals *within* those accounts. All of these thinkers – Aristotle, Cicero, Aquinas and Marx – considered humans and animals to be qualitatively different, and for that reason, they did not consider animals to be appropriate subjects of justice. They may have regarded the treatment of animals to be an interesting private, moral or perhaps religious question; but for these theorists, it was not something which political communities should regulate. To some extent, this approach to the animal issue still lives with political theory today. Many contemporary political theorists regard the question of how we treat animals to be either of no concern, or to be a very trivial concern given the other pressing issues facing political communities today.

But this neglect of the animal issue by political theorists should not be exaggerated. For one, the question of what political communities owe to animals became an important concern for theorists in the 1970s, particularly amongst those in the so-called 'Anglo-American' tradition. As we will see in Chapter 3, Peter Singer's book *Animal Liberation*, first published in 1975, helped to spark a huge interest in the question of how modern communities treat farm, laboratory and other animals. Singer's book exposed the incredible cruelties that contemporary societies inflict on animals for only trivial benefits, and offered a utilitarian framework by which to condemn and remedy such injustices. Importantly, Singer's work elicited, and continues to elicit, a range of responses from political theorists, making questions about whether animals are owed justice, and exactly what they are owed, much more familiar in the contemporary political theory literature.

Furthermore, it is important not to regard engagement with the animal issue solely as a contemporary phenomenon. While Aristotle, Cicero, Aquinas and Marx may have excluded animals from their concerns, other theorists in the history of political thought most certainly did not. Indeed, many of the contemporary debates concerning the question of justice for animals have been addressed before. For this reason, Chapter 2 of the book takes a brief historical overview of the treatment of animals in political thought. The chapter claims that three broad historical trends can be identified: considerable debate and disagreement in Ancient times about whether animals should be included in considerations of justice; consensus in medieval times that animals should be excluded; and then a return of the debate in modern times, especially from the eighteenth and nineteenth centuries. The chapter points out that a number of prominent Ancient thinkers argued that justice should be extended to non-human animals. For example,

Theophrastus, Plutarch and Porphyry all questioned the prevailing assumptions of the day about the killing and eating of animals, and they did so using arguments extremely similar to those offered by contemporary proponents of animal rights. But the chapter goes on to show that these Ancient arguments for animal justice had little influence on the political theorists of medieval Christianity. Instead, the Aristotelian and Stoic idea that only those beings with the capacity for *reason* merit justice proved to be more influential historically. The chapter claims that when medieval Christian writers coupled this focus on reason with *scripture* – scripture which seemed to explicitly sanction man's dominion over animals – it established a consensus in Western political theory that animals are not owed justice. But that consensus did not last forever. As religious orthodoxy waned, and as scientific reasoning grew, a debate about the status of animals began to re-emerge in modern political thought. The chapter shows how the rise of utilitarianism in the nineteenth century was especially influential. After all, and as evidenced by the works of Frances Hutcheson, Humphrey Primatt, Jeremy Bentham and John Stuart Mill, utilitarianism is a theory that is naturally sympathetic to the extension of justice to animals.

Chapter 3 of the book takes up this story by examining utilitarianism's relationship with the question of justice for animals in more detail. It begins by introducing the basic premise of utilitarian thought, explaining that utilitarian thinkers want to devise and recommend policies which aim to *maximise welfare*. Since sentient animals can feel pleasure and pain, and thus very much possess welfares, many utilitarians of the eighteenth and nineteenth centuries were happy to include them in their policy prescriptions. Nevertheless, the chapter claims that animals still remained something of a side issue for these thinkers, with none actually providing a rigorous defence and account of our obligations to animals. This changed in 1975 with the publication of Peter Singer's book, *Animal Liberation*. Because of its influence and because of Singer's explicit utilitarianism, the bulk of Chapter 3 is concerned with examining and evaluating his important work. As such, the chapter briefly outlines the core of Singer's theory, pointing out its implications for such issues as animal experimentation and meat-eating. It then goes on to examine three quite different critiques of Singer. Firstly, it examines the claims of conservative critics who argue that Singer offers too much protection to animals. Secondly, it examines the claims of critics who argue that Singer's policy prescriptions do not logically follow from his utilitarian goal of maximising welfare. And finally, it examines the claims of radical critics who argue that Singer offers insufficient protection to animals. This latter criticism is especially important because it brings up the issue of *animal rights*. To explain, these radical critics argue that Singer's goal of maximising welfare renders his protection of animals ineffective. After all, Singer must in fact support policies that sacrifice individual animals and their interests if those policies maximise overall welfare. As such, these critics claim, in opposition to Singer, that communities ought to protect individual animals and their interests by assigning them *animal rights*.

Chapter 4 explores this tension between utilitarianism and rights-based theories in more detail by examining liberalism's relationship with animals. The chapter claims that the defining feature of liberalism is its valuation of the free and equal individual person. The chapter explains how this ethical core affects both the means by which liberals *determine* what is just, and the *content* they give to justice. To explain, many liberals favour the use of a 'social contract' to devise what a just political community would look like. This is because they believe that it is important that free and equal persons have some say over the political principles that they abide by. Secondly, in terms of the content of those political principles, liberals favour those that support 'pluralism'. This is because liberal thinkers believe that the state should refrain from imposing one particular conception of the good on free and equal persons, but instead allow different ways of life to flourish as far as possible. The chapter explains how these notions of contract and pluralism have been used by liberal thinkers to exclude animals from justice, and even to sanction cruel practices towards them. Interestingly, however, it then goes on to show how some proponents of animal rights have employed these same liberal notions to *extend* justice and strong forms of protection to animals.

Chapter 5 explores communitarianism and its implications for the treatment of animals. The chapter introduces communitarian theory by contrasting it with liberal thought. After all, communitarians take issue with the liberal fetishisation of the non-interfering state, believing that politics is about taking a stand on important moral issues. Communitarians thus argue for a 'politics of the common good', in which the state is unafraid to base policies around the shared moral values of society. Chapter 5 thus considers the possibility of a communitarian theory of justice for animals, in which a society's shared love of animals – as might be claimed to exist in the UK – is used as a basis for awarding extremely strong forms of protection to animals. The chapter goes on to assess a number of key problems with this, and any other, communitarian theory of animal justice. Firstly, there is the problem of what to do about those communities that have no concern for animals. Secondly, there is the grave difficulty in ascertaining any community's 'authentic' shared attitude towards animals. Thirdly, there is the problem that individuals often prefer some animals over others, and often for quite arbitrary reasons. And finally, there is the simple fact that states usually contain a number of distinct communities within them, some of which often have conflicting ideas about the treatment of animals.

In Chapter 6, the Marxist relationship with the question of justice for animals is explored. The chapter acknowledges that identifying a clear Marxist position on how political communities ought to treat animals is extremely difficult. For, on the one hand, Marxism can be viewed as completely *uninterested* in the question of justice for animals. After all, Marxism is not a normative political theory which makes judgements about how communities should run their affairs, but purports to be a scientific theory which describes the conditions for human emancipation. Given this, Marxists might well view the question of how political societies ought to treat animals as entirely irrelevant. However, the chapter points out that Marxism can also be regarded as openly *hostile* to the question of animal justice. Indeed, some Marxist thinkers have accused animal protection movements of representing a 'bourgeois morality' which impedes real socialist concerns relating to the overthrow of oppressive capitalist relations. Nevertheless, and as the chapter explains in its final sections, Marxist resources have also been used by a number of important thinkers in attempts to extend justice to animals. For example, animals have been categorised by some thinkers as an exploited class who are oppressed under capitalism in the same way as workers; 'needs' have been used as a distributive principle by which to extend justice beyond humanity; and socialist insights and methods have been claimed to be necessary for achieving effective political change for animals. This chapter reviews and evaluates all of these issues in detail.

The final theory that the book addresses is feminism, and its implications for the treatment of animals are explored in Chapter 7. The chapter explains that there are various types of feminist theory, and because of this, there is no single definitive feminist perspective on what communities owe to animals. Indeed, the chapter also points out that most feminist writers – just like most utilitarians, liberals, communitarians and Marxists – have had very little to say on the animal issue. Interestingly, however, one particular group of feminist thinkers, so-called 'care-based' feminists, have taken the question of our obligations to animals extremely seriously. Importantly, they have also developed a significant new perspective on the question of how political communities ought to govern their relations with animals. Because of their importance in this regard, Chapter 7 focuses mainly on the work of these care-based feminists. It does so by examining and evaluating three aspects of their work. First of all, it evaluates their claim that the oppressions of animals and women are interrelated, and thus that their liberations are inter-dependent. Secondly, it assesses the important claim of care theorists that traditional reason-based means of extending justice to animals are fundamentally flawed. Finally, it critically evaluates care-based theorists' own proposals for how political communities should govern their relations with animals, proposals which derive from our feelings of care and emotional attachment to animals.

The final chapter of the book concludes by reflecting on political theory and its relationship with animals *as a whole*. In so doing, the chapter makes three broad points. Firstly, it argues that while each of the prominent schools within political theory has its problems, each also has something valuable to offer the debate about how political communities ought to govern their relations with animals. Secondly, the chapter puts forward a brief defence of utilitarianism and liberalism as offering the best resources for communities to draw upon when regulating their relations with animals. It does so by explaining how the wisdom of these theories can be married via an 'interest-based rights approach'. Finally, the chapter and book conclude by making a few tentative remarks regarding the present state of political theory and its engagement with the animal issue. Here it calls for the animal issue to move from the periphery of political theory into the mainstream. For the question of how communities treat animals is far from trivial. In actual fact, given the huge numbers of animals raised, used and killed by political communities, and the enormous suffering inflicted upon them for often the most trivial of benefits, the question of what we owe to animals can coherently be considered as one of the most urgent moral and political issues of our time.

To begin then, the book starts in earnest in the next chapter by examining how this question of justice for animals has been treated in the history of Western political thought.

2
Animals in the History of Political Thought

It would be foolish to jump into an examination of the contemporary debates in political theory about animals without having first understood where those debates have come from. Indeed, it would be quite wrong to consider the animal issue as a purely modern phenomenon which has only exercised political theorists in the last 30 years or so. For in fact, the question of our obligations to animals – whether we have any, and what they are – is as old as political theory itself. Furthermore, contemporary debates about animals are shaped by and often mirror directly debates that have been had in political theory for many centuries. Because of this, this chapter briefly examines the treatment of animals in the history of Western political thought. It does so by organising itself around three periods in Western thought: Ancient, medieval Christian and modern. Inevitably, drawing lines around such periods involves some arbitrariness, as does the selection of the thinkers that is discussed in each. However, structuring the discussion in this way does help to focus on what I take to be an important trend in the discussion of animals in the history of Western political theory. That trend is largely of disagreement about whether animals merit justice in Ancient times, consensus over their exclusion in medieval Christianity and then a return of the disagreement once again in modern times. First of all then, let us turn to the treatment of animals in Ancient political theory.

Animals in Ancient political theory

Most discussions of the ethical treatment of animals in Western philosophy begin with Pythagoras, who was born on the Greek island of Samos, somewhere around 570 BC. We have no surviving texts from Pythagoras himself, nor any texts from contemporaries outlining his particular

philosophical thought. In fact, the evidence we do have for his thinking mainly comes from texts written many centuries after his death.
For example, of particular importance in establishing Pythagoras's reputation was Ovid's *Metamorphoses*, completed around 8 AD.[1] In light
of these facts, there is reason to doubt how accurate our knowledge of
Pythagoras's thought and teaching actually is. Nevertheless, it would be
unwise to entirely dismiss the reports of Pythagoras's life and followers,
especially given the fact that they do largely tell the same tale. That tale
is of a quite remarkable and dexterous thinker. Furthermore, and importantly for our concerns, it also tells of a man and tradition concerned to
show respect to all living creatures.

Pythagoras is famous for his belief in transmigration, or reincarnation.
In other words, he believed that the soul is immortal and can endlessly
be transferred to other living creatures. Given this belief in transmigration, Pythagoras believed that there was a basic kinship amongst all
life, meaning that we should show respect to all living creatures. One
famous story of Pythagoras's life, for example, tells of him stopping an
individual from beating a dog. Pythagoras is said to have stopped the
beating because he recognised in the dog's bark the voice of a deceased
friend.[2] Pythagoras and his followers are also famously associated with
vegetarianism. Indeed, right up until the nineteenth century, when the
term 'vegetarian' was eventually coined, the usual term for a diet that
did not include the flesh of dead animals was the 'Pythagorean diet'.[3]

In light of all this, should we then conclude that Pythagoras offers
us the first account in Western thought of how political communities
should treat animals? No we should not. For although Pythagoras was
proposing a certain form of behaviour with regards to animals, we cannot say that he was ever putting forward a set of rules for political
communities to follow. For it is important to re-emphasise that political theory is not concerned with each and every question about how
we should live, but rather, is concerned with a particular subset of those
questions. As was stated in the introduction to this book, political theory
involves the investigation of how *political communities* should govern
themselves; that is to say, political theory is concerned with the question
of *justice*. Without doubt, this question of justice, of how political communities should govern themselves, is not what concerned Pythagoras.
Of course, he may have thought that individuals ought to refrain from
eating meat, and that they ought to refrain from being cruel to animals,
on top of many other aspects of his teaching. But did he actually believe
that political communities should permit, forbid or require individuals
to behave in these ways in relation to animals? That is highly unlikely.

Pythagoras has mischievously been described by Bertrand Russell '... as a combination of Einstein and Mrs Eddy'.[4] What Russell means by this is that Pythagoras was both pure mathematician and religious prophet. His teachings on how to behave, only a very small part of which related to animals, were ascetic practices, designed for his followers to use in order to purify their souls. Pythagoras may have been moralising – preaching even – but he was not outlining coercive rules intended to govern us in political communities. This is in no way a criticism of Pythagoras, it is simply to point out that what he was doing was not political theory.

However, a clear example of someone who did do political theory, and who did have justice as his focus, was Plato (427–347 BC), who lived around a century after Pythagoras. While Pythagoras is often regarded as the starting point for thinking about animal ethics in Western thought, Plato is often regarded as the starting point for thinking about Western political theory. In his famous work, *The Republic*, Plato outlines what he takes justice to be – what the ultimate political community should look like.[5] He is not merely moralising about what he thinks individuals ought to do in order to be decent and honourable people; rather, he is telling us what they are permitted, forbidden and required to do to conform to justice. As mentioned above, it is outlining what justice requires that continues to be the primary concern of political theorists to this day. For Plato, justice does not simply correspond to the customs of a society, nor does it derive from what members of a society at any given time might prefer. Rather, for Plato, we can arrive at an objective understanding of what justice is by using the power of reason. Importantly, reason not only provides us with the method to discover justice, but it also provides us with an important part of the *content* of justice. To explain, for Plato, social justice consists in individuals acting in accordance with roles related to their natural aptitudes. This involves a highly structured society in which some produce, some protect and others rule. For a fully ordered society, Plato claimed that reason must ultimately rule, and hence his Republic is famously governed by philosopher-kings. These philosopher-kings do not just get to rule because they are naturally the cleverest however. Instead, each goes through a strict system of education and socialisation in order to be able to understand what justice actually is, and thus rule in accordance with what is just.

What then of Plato's treatment of animals? It has to be noted that *The Republic* does not tell us whether justice pertains to animals. It is certainly true that Plato does speak in the book of an ideal condition in which humans do not eat animals.[6] And in fact, some thinkers have taken this to imply that Plato's ideal state would be vegetarian.[7] However, we should be cautious here. After all, Plato is quite explicit later in the book that a more luxurious state is permissible, and that meat-eating would be allowed in such a society.[8] Furthermore, since Plato states that the eating of meat would require this more luxurious society to have more doctors, we can infer that his ideal preference for vegetarianism is based more on reasons of health than any putative obligations to individual animals themselves.[9]

So while Plato may have been the first to offer a detailed account of justice in Western political thought, he said next to nothing on the question of justice for animals. However, Plato is hugely important to our concerns because of the emphasis he gave to reason. Plato made reason absolutely central to the question of justice: it was the method of discovering what it is, and it provided a crucial element of its content. What Plato did not do, however, was explicitly state that the capacity for reason is necessary in order to benefit from justice. This claim was in fact first made by Aristotle (384–322 BC). It is in Aristotle's work that we find the first direct engagement with the question of justice for animals, and it is in his work that we find the most influential grounds for excluding them: their irrationality. The argument that animals lie outside of justice because they lack reason was hugely important in subsequent Western political theory, and continues to be repeated by several thinkers up to this day. As we will see, the claim was repeated in Stoic thought, and then repeated once again by Christian thinkers. The influence of these two schools of thought on the Roman and medieval worlds is obvious, and goes some way to explaining the longevity of the exclusion of animals in Western political theory. However, it was Aristotle who first spelt out the argument that animals do not merit justice because they are irrational, and so it is worth looking at his reasoning more closely.

For Aristotle, all natural things have an end or purpose, what he calls a *telos*. Given that man (and he does mean man) is part of nature, he also has an end. Because of man's unique moral and intellectual capacities, Aristotle believed that man can only reach full development of his end communally within the *polis*. Man, famously for Aristotle, is a *politikon zōon*: a 'political animal'.[10] What Aristotle means by this is that the full development of man's moral and intellectual nature can only be achieved within this political association. Ruling, being ruled, debating on the rights and wrongs of society and so on are what it means to participate as a citizen, and what it means to be a fully flourishing individual. As such, the political is a domain purely for men. Politics requires the ability for rational thought, the ability to reflect on and pursue the virtuous life:

> For the real difference between man and other animals is that humans alone have perception of good and evil, just and unjust etc. It is the sharing of a common view in *these* matters that makes a household and a state.[11]

For Aristotle, then, the distinction between men and animals is fundamental. Justice requires a certain affinity or commonality between individuals. Because men are rational, and animals are not, no such affinity exists between men and animals, according to Aristotle.[12]

Of course, Aristotle was concerned with how the capacity for reason can found a sense of community within the *polis*: the independent city-state. Similarly, Plato too was interested in the rule of reason within a *polis*. However, Zeno of Citium (*c.* 335 BC–*c.* 263 BC), the founder of the Stoic school of philosophy, wrote a *Republic,* which is now lost, which spoke of a *worldwide* city in which all men are citizens. Stoics like Zeno propounded the notion of a 'brotherhood of man' which was founded on their common rationality. This humanistic universalism has been incredibly important in the history of political thought. For not only does Stoicism itself extend into Roman thought, through the writings of Cicero (106 BC–43 BC), Seneca (4 BC–AD 65) and Marcus Aurelius (AD 121–180), but it also influenced Christianity, and even modern notions of human rights. Of key importance for Stoics was the idea of a universal natural law which can be accessed by humans through reason. Importantly for our purposes, because reason is required to access this universal law, Stoics claimed that it is only human beings who are regulated by and benefit from that law. Indeed, Chryssipus (*c.* 290 BC–*c.* 206 BC), the third head of the Stoic school in Athens, was quite explicit in denying justice to animals on the simple basis that they are not rational.[13]

In the hugely influential writings of Plato, Aristotle and the Stoics then, we see the establishment of an important connection between reason and justice; a connection that survives in political theory to this day. Moreover, it is a connection which has often served to exclude animals from the domain of justice, and to keep our treatment of animals off the radar of political theorists. But we should not assume that all Ancient thinkers excluded animals in this way. There was a range of important Ancient thinkers who resisted these Aristotelian and Stoic claims, and argued that justice could and should be extended to animals. Really, the existence of such thinkers should come as little surprise. After all, in order for the reason-based exclusion of animals to work, two things need to be shown: firstly, that the capacity for reason is a necessary prerequisite for receiving justice; and secondly, that animals do lack reason. Quite obviously both of these claims can be denied, and this was evident enough to many Ancient philosophers.

For example, Empedocles (*c.* 495 BC–*c.* 435 BC), claimed that justice could be extended to animals because we have a kinship with them. Like Pythagoras, Empedocles believed in reincarnation, and so also believed that killing animals for food or for sacrifice literally involves slaughtering our kin. Importantly, for Empedocles the prohibition on killing animals was not based on ascetism, but was a law of nature and thus a matter of justice.[14] Aristotle's student and successor at the Lyceum, Theophrastus (*c.* 372 BC–*c.* 287 BC), also believed that killing animals was unjust. However, he claimed that killing animals was unjust simply because it robbed them of life. Clearly then, we can see that for some important Ancient thinkers, the capacity for rationality was not the crucial determinant of whether one could be treated justly: for Empedocles it was kinship; and for Theophrastus it was the possession of life itself.

Interestingly, a number of Ancient thinkers also challenged the idea that animals are irrational. Perhaps the most famous of these is Plutarch (*c.* 45 AD–*c.* 120), whose essay 'On the Cleverness of Animals' directly challenged the Stoic denial of reason to animals. While the Stoics argued that animals provide for themselves and keep alive purely through some kind of unreflective instinct, Plutarch pointed out that the act of desiring and seeking out goods for survival surely involves *some* level of reasoning.[15] It may not be exactly the same kind of reason that we as humans employ, nor necessarily of the same level of sophistication, but it can plausibly be considered as reasoning nonetheless.[16] The attribution of such reasoning to animals allowed Plutarch to maintain that we do have duties of justice to animals, but it also led him to go on to consider *what* duties we have to them. Interestingly, Plutarch claimed that we have no obligation to abstain from killing those animals that harm us. Of particular fascination, however, is his argument that it is permissible to domesticate those that are useful to us.[17] Here, then, is one of the earliest discussions of whether justice for animals involves the abolition of all animal *use*, something that is debated fiercely by contemporary theorists.[18]

Plutarch was a major influence on Porphyry (232 AD–309 AD), whose book, *On Abstinence from Animal Food,* must surely be the most complete and compelling defence of justice for animals in Antiquity. Porphyry, like Plutarch, also attributed rationality to animals, and this forms part of his multi-faceted defence of extending justice to them. One of the most interesting passages in the book, however, concerns not rationality, but sentience. Porphyry tackled the question – all too familiar to contemporary vegetarians – about the difference between killing animals and killing plants. For if killing animals for food is prohibited, why not also prohibit the killing of plants? Porphyry's response was simple enough, but remains incredibly important:

> To compare plants, however, with animals, is doing violence to the order of things. For the latter are naturally sensitive, and adapted to feel pain, to be terrified and hurt; on which account they may also be injured. But the former are entirely destitute of sensation, and in consequence of this, nothing foreign, or evil, or hurtful, or injurious, can befall them. For sensation is the principle of all alliance....[19]

In this paragraph Porphyry claims that it is sentience – the capacity to experience the world – that is crucial for being wronged and receiving justice. As we will see, this claim was repeated some 1500 years or so later, by the utilitarian political theorist Jeremy Bentham. Moreover, sentience provides the reason why many contemporary political theorists consider the treatment of animals to be a matter of justice.

Clearly then, it would be quite wrong to conclude that the story of Ancient thinkers' consideration of animals is simply one of exclusion. As we have seen, there is no single Ancient perspective regarding what political communities owe to animals, but rather, a range of sophisticated and different views. Indeed, it is quite remarkable how many contemporary debates about animals are debates that were previously had in Ancient Greece and Rome. Nevertheless, it is also clear that some Ancient attitudes to animals have been more influential than others. Indeed, it has been the Aristotelian and Stoic denial of reason and justice to animals that has had the greatest impact on the history of Western political thought. This is particularly evident in the writings of the most important Christian political theorists, which we examine in the next section.

Animals in medieval Christian political theory

The two most important and influential thinkers in the history of Christian political thought are usually taken to be Augustine (354 AD–430 AD) and Thomas Aquinas (*c.* 1225–1274). Given the importance of these figures – and of course the authority of the Church in the medieval period – the influence of their teachings in relation to our treatment of animals is hard to overstate. Importantly, both were explicit in their denial of justice to animals. Augustine claimed that we have no ties of justice to animals because they lack rational souls. For Augustine, animals have been put on Earth by God for us to kill and keep alive as we wish.[20] Aquinas repeats these views a millennium later. However, he goes even further than Augustine. Not only does Aquinas deny that we have duties of justice to animals, but he also denies that we owe any duties of charity to them.[21] For Aquinas, animals exist on Earth for human use, and the obligation to refrain from treating them cruelly rests entirely on the fact that such actions might lead us to treat other humans cruelly.[22] What we see in both thinkers, then, is the entrenchment into Western political thought of two positions: firstly, a repeat of the Aristotelian and Stoic claim that the possession of human reason is a necessary prerequisite for receiving justice; and secondly, the hugely influential idea that any limited duties we do have in respect of animals are not owed directly to animals themselves, but are really owed to mankind.[23]

To some, the teachings of these Christian writers about our proper treatment of animals might not be much of a surprise. After all, it is commonly said that what makes the Western attitude to nature distinctive is that human beings are not regarded as elements of the natural world, but instead as things which are external to and superior to it.[24] It is also commonly stated that the foundation for this view derives from the Judeo-Christian tradition, and is sanctioned directly in the book of Genesis.[25] Indeed, Genesis 1: 26 famously states:

> Then God said, 'Let us make man in our image, in our likeness, and let them rule over the fish of the sea and the birds of the air, over the livestock, over all the earth, and over all the creatures that move along the ground.'

In this passage we hear that human beings are special and stand apart from the rest of nature in that they alone are made in the image of God. And we also hear quite explicitly that human beings are entitled to rule over the rest of the natural world. The positions of Augustine and Aquinas, it might be argued then, make perfect sense. Of course humans can do what they like to animals, because that position is directly sanctioned by the story of Creation itself.

However, it would be quite wrong to think that Augustine and Aquinas derived their attitudes to the treatment of animals simply by looking up the relevant passages of the Bible. After all, what does 'rule over' the animals actually encompass? It certainly does not necessarily entail complete license to treat animals however we please. Benign paternalism is as much a form of rule as unconstrained tyranny.[26] Indeed, it is obviously possible for one group to rule over another, and for justice to exist between them. As such, we should be wary of putting the positions of Augustine and Aquinas solely down to scripture.

For Augustine, the denial of justice to animals derives from two main thoughts. First of all, he draws on the Stoic idea that we have no moral obligations to animals because they lack reason. He argues that we can kill plants because they have no sensation, and we can kill animals because they have no 'rational community' with us.[27] Secondly, it is unclear how interested Augustine was in the idea of political justice between *human beings*, let alone between humans and animals. After all, according to Augustine, the state is merely a necessary evil to keep our sinful natures in check. All it can do is keep the peace, it certainly cannot achieve the kind of utopia imagined by Plato, nor can it lead men to the enjoyment of fully flourishing lives as envisaged by Aristotle. Because of the Fall, and because of humanity's inherently sinful nature, Augustine believed that true justice cannot be attained in any Earthly political organisation. For Augustine, true justice can only be attained in the City of God, his metaphor for heavenly utopia. As such, Augustine was not greatly concerned with the precise details of how political communities should govern themselves. His lack of concern for how political communities treat animals thus derives in part from a scepticism about the very idea of political justice itself.

For Thomas Aquinas, on the other hand, the state can and should do more than simply act as a check on sin. Synthesising Aristotelianism with Christianity, Aquinas argued that the state has the purpose of directing men to two ends. First of all, and in contrast to Augustine, Aquinas believed that the state can lead men to a just and moral life here on Earth. Secondly, in helping men lead a moral life, the state can also facilitate the achievement of salvation. Of course, this second end cannot be achieved solely by the state; salvation also requires the active participation of the Church. Now for Aquinas, what is crucial for the achievement of these ends is the exercise of reason. According to Aquinas, Earthly justice and salvation can only be achieved by acting in accordance with the law. The law is not simply the customs of one's society, nor is it that which is most expedient to keeping the peace. Rather, it is the natural law of God, attainable to men through the exercise of reason. Clearly then, Aquinas is not borrowing only from Aristotle in his philosophy: there is an important Stoic, and even Platonic, theme recurring here. And like the Stoics, Aquinas sees reason as absolutely central to the question of who merits justice. In fact, and as mentioned above, Aquinas also sees the possession of reason as crucial to determining who merits mere *charity*:

> ...the love of charity extends to none but God and our neighbour. But the word neighbour cannot be extended to irrational creatures, since they have no fellowship with man in the rational life. Therefore charity does not extend to irrational creatures.[28]

By taking Biblical ideas about the uniqueness of man and man's dominion over the Earth, and fusing them with the Aristotelian and Stoic emphasis on reason, Augustine and Aquinas excluded animals from justice in Christian political thought. Nevertheless, while their views on the treatment of animals were highly influential, we do need to consider whether they were shared by *all* medieval Christian thinkers. After all, St Francis of Assisi (*c.* 1181–1226), the founder of the Franciscan order, is popularly known as the patron saint of animals. He is famous for his compassion to animals, and is even said to have preached to birds and other animals. Does St Francis thus represent the existence of a debate about the question of justice for animals in medieval Christianity; a debate reminiscent of that which existed in the Ancient world? Unfortunately, he does not. For one, we have little evidence outlining precisely what St Francis thought and preached. Francis's own writings which do survive barely mention animals, and focus on more orthodox questions of the time such as poverty, chastity and humility. Indeed, Francis's reputation as a lover of nature largely derives from the book, *Little Flowers of Saint Francis*, which was written by a Franciscan a century after Francis's own death.[29] Secondly, even if we take seriously these later accounts, there is still no evidence that Francis was interested in the question of *justice* for animals. He may have had compassion for animals, but he equally had such affection for flowers, the Moon, the Sun, water and fire, as elements of God's creation.[30] Importantly, this concern to praise all of God's creation never amounts to a discussion of how political communities should regulate their treatment of animals. And as we have seen, it is this question of what counts as justice that is the primary concern of political theorists.

So it is safe to say that political theory in medieval times did not contain the sort of debate about the question of justice for animals that existed in Ancient times. Augustine and Aquinas helped to establish the orthodoxy of the time by fusing Ancient philosophical arguments with scripture. They took the Aristotelian and Stoic focus on reason, combined it with the Biblical sanctioning of man's dominion, and established the grounds for excluding animals from justice. Given the importance of these thinkers to the Church's thinking and teaching, and given the Church's authority in social and political matters in medieval times, it is little wonder that such views held sway for such a long period. However, what is fascinating is how influential these arguments continued to be when the importance of religious orthodoxy waned, and when the debate over justice for animals reawakened. It is the discussion of this reawakening that is examined in the next section.

Animals in modern political thought

The Prince by Niccolò Machiavelli (1469–1527) is usually considered to be one of the first works of 'modern' political thought.[31] While it is not hugely significant in terms of what it has to say about animals, it is significant in the way it addresses political theory. *The Prince* is considered to be modern and novel because of its focus on men and politics as they are, as opposed to what they might be. In other words, its search for improvement starts with what we have and what is possible, rather than with lofty ideas about Earthly or spiritual utopias. Of course, this break with previous ways of doing political theory can be overstated – there are clear comparisons with the methodology of Aristotle and Augustine here – but nevertheless, Machiavelli is considered to be novel in taking this more scientific approach to the subject of how political communities should govern themselves. The purpose of *The Prince* is to advise new rulers on the establishment and consolidation of their rule. The essential proposition is to do whatever it takes for the maintenance of order: to maintain popular support; to be loved and feared; to be cruel if necessary; to appear pious if necessary; and above all to adapt to circumstances. Justice in *The Prince* is equivalent to whatever it takes to achieve order and stability. For it is only in an ordered and healthy state that good laws can be established. So while the book tells us next to nothing in terms of how political communities should regulate their treatment of animals, it is clear that Machiavelli's answer would simply depend on how it helps the Prince achieve harmony in society. For Machiavelli, animals should not necessarily be treated badly, nor should they necessarily be treated well. Instead, everything depends on the circumstances of the time and place.

This modern turn in philosophy had some important implications for the way individuals of the time thought about animals. Indeed, the French philosopher René Descartes (1596–1650), often dubbed the father of modern philosophy, had some notoriously controversial views on animals. Descartes believed that the physical world could be understood mechanistically, and that this applied to human beings as much as anything else. As such, he argued that the physical aspect of human beings – the human body – could be understood as working like a machine. Our minds, on the other hand – how and why we think, feel and speak – cannot be so understood. For Descartes, the working of our minds does not derive from any material processes, but from the fact that we have an immaterial soul. In other words, we think, feel and speak because we possess souls. Descartes noted that animals are essentially made of the same stuff as us: blood, tissues, organs and so on. As such, their bodily functions can be explained mechanistically in the same ways as our own. At the same time, however, quite clearly animals cannot speak. This was easily explained by Descartes by the fact that they lack immortal souls, something which of course was hardly controversial at the time. Because they lacked such souls, however, Descartes also thought that they had no capacities of the mind at all: no speech, no reason and no feeling either. For Descartes then, animals were completely material, mechanistic entities, with no consciousness whatsoever.[32] Some thinkers have made a direct link between this Cartesian position on animal sentience and the horrific experiments on live animals that occurred throughout the seventeenth and eighteenth centuries. For in that period, animals were routinely cut open while alive for scientific enquiry, and sometimes they were cut open in public displays. If anyone was naïve enough to enquire what was causing the screams of the supposedly non-sentient animals involved, it was pointed out that these were not screams of pain, but simply unfelt mechanical responses.

Descartes was a hugely important philosopher and it would come as little surprise if his views on the capacities of animals had been seized upon by large numbers of his contemporaries in order to deny justice to animals. And yet, that seems not to have been the case. Many were sceptical of Descartes's denial of sensation to animals, perhaps because it was just too much of an affront to common sense. Many even opposed the vivisection practices described above. Nevertheless, the dominant position of political theorists in the seventeenth and eighteenth centuries was still to exclude animals from justice. Interestingly, however, that exclusion was on the familiar basis that animals lack reason. Reason became – or rather remained – crucial to political theorists because of the rise of so-called 'contract theories'. We will return to contract theories in Chapter 4 when we examine contemporary liberal political theory in detail, but it is worth looking at the origins of the idea and how it impacted on the question of justice for animals. The most obvious place to start that discussion is with the work of Thomas Hobbes (1588–1679).

For Hobbes, the content of justice depends simply on the ordinances laid down and enforced by the rulers of a political community. Justice is not a set of universal rules determined by the application of reason, but is simply what the political authority says it is. For Hobbes, we need such rules, and should obey such rules, because of the immense benefits of political authority. Hobbes calls life without political authority the 'state of nature', and depicts it is a savage war of all against all. Hobbes thus follows the modern idea of taking men as they are, and takes them to be naturally brutal. In order to overcome the fear and insecurity of this horribly violent state of nature, Hobbes argues that it is rational and prudent for individuals to make a contract with one another. In that contract, individuals give up all of their natural rights and liberties to an ultimate political authority. That political authority makes the rules, and those rules should be obeyed in order to avoid slipping back into the chaos of the state of nature.[33] For our purposes, what is crucial about this story is Hobbes's claim that animals stand outside of this contractual process because of their irrationality. Animals cannot participate in the contract, and so for Hobbes stand outside of the political community, and outside of the domain of justice. In Hobbes's terms, because we cannot make a contract with them, we are in a perpetual state of war with animals.[34]

This notion of a social contract was later used in very different ways by such thinkers as John Locke (1632–1704) and Jean-Jacques Rousseau (1712–1788). While each of their theories is importantly different from Hobbes's and from each other's, the use of a contract to determine justice served to exclude animals in the same way. Contract theorists ask what individuals setting up a political community would agree to: what they would give up to that community, and in return for what. For Hobbes, as we have seen, individuals would choose to give up all of their natural freedoms in return for security. For Locke, individuals would give up their natural right to punish others, in order to enjoy their remaining rights more securely.[35] For Rousseau, individuals would submit themselves entirely to a sovereign authority, in order to live a truly free life as a citizen governing through the general will.[36] Importantly for our concerns, in each case, individuals choose to constrain their behaviour in some way in order to receive something more valuable in return. Of course, and as all of these thinkers accepted, whatever their capacities, it is clear that animals cannot participate in this kind of reasoning: they cannot choose to give something up in return for something greater. As such, animals remain outside of the contracting process, and thus outside of the domain of justice. While some contemporary thinkers have attempted to reconcile contractarianism with the extension of justice to animals, the legacy of contract-based political theory for animals has primarily been exclusion.

Of course, the reason why animals cannot participate in the formation of any contract is not so much because they cannot reason at all, but rather, that they cannot perform a particular type of reasoning. Indeed, Rousseau himself argued that the fundamental distinction between humans and animals is not the capacity for sensation, or reason, but the capacity for *free will*.[37] And it is surely the capacity for ethical *reflection* and *choice* – or 'autonomy' – which seems crucial for the formation of a contract. However, the thinker most famously associated with the prizing of autonomy is not Rousseau, but rather the philosopher Immanuel Kant (1724–1804). Kant claimed that human beings possess dignity. They possess dignity, he argued, because they are autonomous beings, or *ends-in-themselves*: self-determining entities who can exercise moral judgement and free will. Because they are ends-in-themselves, Kant argued that possessors of dignity should always be treated as ends, and never solely as a means. Essentially this means that individuals should be free as far as possible to reflect on, decide and pursue their own goals in life. Those entities who cannot exercise moral judgement and free will, on the other hand, do not possess dignity, are not ends-in-themselves, but are instead mere 'things', who can permissibly be treated as mere means. Importantly for our concerns, Kant argued that animals belong in this class of things. That does not mean to say, however, that Kant believed that there are no limits to what we can do to animals. We should refrain from cruelty towards animals, he argued, and we should show them compassion. Importantly, however, and directly mirroring the arguments of Aquinas before him, Kant argued that we should do so only because 'Tender feelings towards dumb animals develop humane feelings towards mankind.'[38] In other words then, there are no direct duties to animals themselves, merely indirect ones that are really owed to mankind.

Kant's legacy should not be underestimated. By hanging so much moral weight on the autonomy of human beings, Kant provided a secular justification for the dignity of humans. Humans are special and have dignity, according to Kant, not because they possess some spark of Godliness or immaterial soul, but because they are self-governing entities. Such claims have been extremely influential in contemporary politics. For example, they can clearly be seen in the human rights movement. Indeed, the notion of 'human dignity' has been used to provide a basis for assigning universal rights to each and every human being on the planet without resort to contentious religious claims. Given the central place of human rights in contemporary debates about justice, Kant's contribution to political theory thus remains incredibly important. And it remains important for animals too, quite simply because it excludes them. In fact, the argument is still regularly heard that since animals are not autonomous creatures, they do not possess dignity, and thus are not appropriate bearers of rights.[39]

But in spite of the thoughts and legacies of Descartes, contract theorists and Kant, it would be quite wrong to conclude that all strands of modern political thought have served to exclude animals from justice. An extremely important line of thought in modern political theory emerged in the eighteenth and nineteenth centuries which was very sympathetic to the plight of animals. Utilitarianism, which has the maximisation of welfare as its primary concern, re-emphasised the link between sentience and justice. I say 're-emphasised' because, as we saw above, Porphyry made this connection way back in the third century. Nevertheless, from the eighteenth century a number of utilitarian thinkers claimed once again that animals merit justice because of their capacity to feel pleasure and pain. In other words, they claimed that when formulating policies to increase the welfare of a society, the interests of animals should be included. The Scottish philosopher Frances Hutcheson (1694–1746) was perhaps the first modern thinker to explicitly ground animal rights in their capacity for pleasure and pain, and hence in their contribution to society's overall happiness.[40] A Christian utilitarian basis for including animals in the domain of justice was also laid out in detail in Humphrey Primatt's (*bap.* 1735, *d.* 1776/7) book, *The Duty of Mercy and the Sin of Cruelty to Brute Animals.*[41] However, the most famous case for bringing animals under the domain of justice on the basis of their sentience derives from Jeremy Bentham (1748–1832). In a famous passage linking sentience to justice, Bentham wrote as follows:

> The day *may* come, when the rest of the animal creation may acquire those rights which never could have been withholden from them but by the hand of tyranny. The French have already discovered that

the blackness of the skin is no reason why a human being should be abandoned without redress to the caprice of a tormentor. It may one day come to be recognized that the number of the legs, the villosity of the skin, or the termination of the *os sacrum* are reasons equally insufficient for abandoning a sensitive being to the same fate. What else is it that should trace the insuperable line? Is it the faculty of reason, or, perhaps, the faculty of discourse? But a full-grown horse or dog, is beyond comparison a more rational, as well as a more conversable animal, than an infant of a day, or a week, or even a month, old. But suppose the case were otherwise, what would it avail? the question is not, Can they *reason*? nor, Can they *talk*? but, Can they *suffer*?[42]

Bentham thus separated rationality and autonomy from the question of who merits justice. In the passage above, he points to the example of small infants and asks why it is that we include them in our moral and political considerations. We do not include them because they are rational or autonomous, he argues, for obviously they are not. Rather, we include them, he claims, because they can *suffer*, because they are *sentient*. Bentham claims that it is this ability to experience the world that makes an entity deserving of justice.

Bentham's focus on sentience is important in and of itself, but also because of its influence on political theory more generally. For one, the inclusion of animals on the grounds of their sentience was supported by John Stuart Mill (1806–73), who had an enormous impact on both contemporary utilitarian and liberal thought. Secondly, Bentham's ideas also formed the basis of Peter Singer's 1975 book, *Animal Liberation*, which helped to make our treatment of animals a prominent issue in mainstream political discourse. In all, it is fair to say that the rise of utilitarian thought in the eighteenth and nineteenth centuries also led to a rejuvenation of the debate concerning what political communities owe to animals.

It will not do, however, to leave the discussion of modern political theory there. For given his enduring influence on political theory, any discussion of modern political thought ought to make some mention of Karl Marx (1818–83). We will discuss Marxist attitudes to animals in full in Chapter 6, but it is worth introducing Marx's own thoughts briefly here. Marx held the kind of utilitarian theory described in the paragraphs above with utter contempt. Attempting to reform society so that it better promotes utility was, for Marx, a futile task. According to Marx, human beings can never truly be 'happy' while they live under the oppressive relations of capitalist production. As such, attempting to adjust the system through utility-promoting legislation is counter-productive. Moreover, attempting to improve the lot of animals by including them in such legislation is similarly pointless. For Marx, history is about the unfolding of *human* productive forces. The end point of that process is communism, and Marx claimed that under communism human beings will overcome scarcity and thus be able to live truly free and human lives. Accordingly, Marx claimed that political theorists should spend less time concerned with what the just policies and institutions of contemporary society might look like, and spend more time exposing its inherently oppressive relations. As such, Marx would have thought that the question of what political communities owe to animals to be entirely irrelevant.

From the preceding discussion then, it should hopefully be clear that the emergence of modern political thought did not herald any kind of consensus about the treatment of animals. Nevertheless, it did herald a return of a *debate* about their treatment. To be sure, most political theorists still excluded animals from the domain of justice; and as we have seen, some of the most influential figures in modern political thought – Descartes, Kant and Marx for instance – left legacies which have served to maintain that exclusion. Nevertheless, such positions were challenged, and challenged by important figures in political theory. By the eighteenth and nineteenth centuries, the question of what political communities owe to animals was not considered absurd. Influential and important thinkers such as Hutcheson, Bentham and John Stuart Mill all championed the idea that political communities have obligations to animals. In reality, this concern should be of little surprise. After all, political theorists are shaped by the social attitudes and movements of their time. Importantly for our concerns, the nineteenth century was the period when many political communities in the West began to regulate their conduct towards animals for the first time: setting up societies for their protection, and passing legislation to outlaw cruelty towards them. It is little wonder then, that some of the political theorists of the time provided theoretical underpinnings to such movements.

Conclusion

As pointed out at the start of this chapter, the discussion of animals in the history of Western political thought has ebbed and flowed. Whether justice extends to animals was a matter of important consideration in

Ancient times, and the disagreements between philosophers of the time were quite pronounced. Nevertheless, some of these arguments were more influential than others. And it was the Aristotelian and Stoic idea that animals do not merit justice because they lack reason which had the greatest impact on Christian writers in medieval times. Authors such as Augustine and Aquinas fused this focus on reason with the Biblical sanctioning of man's dominion over the animals. This fusion served to exclude animals from the consideration of political thought for over a millennium. However, with the collapse of orthodoxy also came the development of more modern and scientific thinking. Some of this scientific thought served animals badly: Descartes's denial of sentience to animals is certainly significant, but perhaps more important is the rise of the contractarian method of determining justice. And yet, some modern thinking also served animals rather well, and sought to extend justice to animals. Of particular significance is the re-emergence of the importance of sentience amongst utilitarian political theorists of the nineteenth century.

But while this story is undoubtedly interesting in and of itself, it also has a wider significance for this book. For this story helps us to think more clearly about current debates about animals amongst contemporary political theorists. For one, many of these contemporary debates have been addressed before, and it would be foolish to ignore past wisdom. Furthermore, even debates which were not directly addressed in the past have still been shaped by attitudes to animals that derive from this historical context. A basic understanding of that context should thus serve us well in future chapters. For example, Chapter 3 addresses utilitarianism's relationship with the issue of animal justice, and explores debates about which policies would best promote the welfare of animals. The obvious starting place for any such discussion is of course the work of Bentham; but it would be wise too not to forget the important contributions of Plutarch and Porphyry in this regard. Chapter 4 addresses liberalism's treatment of the animal issue. Here we will see that most of the contemporary debates about animals within liberal political theory centre around the issues of the social contract and personhood; issues that, as we have seen, date back to the beginning of political theory itself. Chapter 5 examines communitarianism, which is a relatively recent school of thought in political theory. However, communitarianism draws on notions of kinship that were of crucial importance to the Stoics, and perhaps even to Pythagoras. Chapter 6 looks at Marxist attitudes to animals, and so of course must address the work and legacy of the great thinker himself. And finally, Chapter 7 addresses feminism's relationship with the question of justice for animals. Here we will address a particular strand of feminist thought named care-based theory, which is extremely sceptical about the centrality of *reason* in political theory; a centrality which as we have seen was entrenched in Western political theory from its very beginnings.

With luck then, because of this historical discussion, we are better placed to embark on an examination of the debates about animals in contemporary political theory. That examination begins in earnest in Chapter 3 which explores utilitarianism and its implications for the question of how political communities should govern their relations with animals.

3
Utilitarianism and Animals

In the previous chapter we heard how a number of eighteenth and nineteenth century thinkers proposed that justice extends to animals. Thinkers such as Frances Hutcheson, Humphrey Primatt, Jeremy Bentham and John Stuart Mill all claimed that political communities owe something to animals, and owe them something on the basis that they are sentient. While such claims were not entirely novel, they had certainly not been heard for some time. The dominant position in medieval times had been that animals are fundamentally different to humans because they lack reason; and such a claim had spilled over into modern times, with many thinkers arguing that animals are owed nothing because of their lack of free will and their inability to participate in a social contract. However, the emergence of utilitarianism as a discrete political theory in the eighteenth and nineteenth centuries posed a radical challenge to such views, and allowed the connection between sentience and justice to be made again.

One useful way of understanding utilitarianism is to see it as the epitome of *modern* political theory. As we saw in the previous chapter, what is said to mark out the modern period of political thought is the wane in influence of religious orthodoxy, and the rise in influence of scientific method. Utilitarianism is certainly modern in both of these senses. After all, utilitarianism takes as its measure of right and wrong, of good and bad legislation, and of just and unjust institutions, not whether those actions, policies or institutions conform to God's will or to some religious ordinance, but whether they promote utility. Moreover, utility is neither divine, ethereal, nor obtained from musings about hypothetical social contracts; instead utility is something real and which can be measured scientifically. For example, for classical utilitarians like Jeremy Bentham, utility can be equated with pleasure, and as such it is the obligation of political communities to formulate policies and institutions which promote pleasure. Utilitarianism then, has no respect for tradition, custom, orthodoxy or received wisdom per se. All actions, policies and social norms are judged in terms of their promotion of utility, and this can lead to radical criticism of existing political structures, and radical proposals for their remedy. Indeed, historically utilitarianism has been at the forefront of social reform. Figures like Bentham, Mill and Singer have not been mere armchair philosophers, but have been staunch campaigners against slavery, racism and poverty, and equally staunch proponents of public health, public education, democracy and women's rights. It is worth remembering that all such reform movements were once perceived as dangerously radical, conflicting as they did with ancient norms and laws. However, as stated above, utilitarianism is unafraid to sanction policies that lead to radical societal transformation, so long as they serve to promote utility.

Returning to the issue of our treatment of animals then, it should hopefully be clearer why so many utilitarian thinkers have promoted the extension of justice to animals, in spite of the relative unorthodoxy of such claims. In ethical terms, utilitarianism is both *welfarist* and *egalitarian*. It is the combination of these two features which makes the promotion of justice for animals so natural for utilitarians. Utilitarianism is welfarist in that it defines what is good solely in terms of welfare. Welfare may go by a variety of different names – utility, well-being, interest-satisfaction and so on – but it is the promotion of this single good that is of ethical importance to utilitarian theories. Now, while it is true that utilitarians differ as to what utility actually entails, most agree that animals can possess it. For example, classical utilitarians, like Bentham, equate utility with pleasure, the capacity for which animals certainly seem to possess. Preference utilitarians, like Singer, equate utility with preference satisfaction, the capacity for which animals also obviously seem to possess.

However, while animals may have the capacity for welfare, why should political communities include them in their utility calculations? Why don't they just ignore the welfare of animals, and seek to promote only the utility of human beings? The reason utilitarians do not generally recommend this is because of the *egalitarian* nature of the theory. Utilitarianism is not egalitarian in the sense that it believes that everyone and everything should receive the same treatment or possess the same set of entitlements. But it does believe that like cases should be treated alike. Thus if the ultimate political goal is to promote welfare, then all entities whose welfare can be promoted must be included in that goal. Just as it is arbitrary to exclude from our policy formulations the welfare of those of a different race, gender or class, so too is it arbitrary to exclude those of a different species. Welfarism defines utility as the sole measure of right and wrong, and egalitarianism demands that all beings with a welfare be included in a political community's pursuit of utility. As such, utilitarianism can extend justice to animals with relative ease.

In spite of these facts, one should not get the impression that all utilitarians from Bentham onwards have been tirelessly championing the cause of animals. It must be remembered that Bentham's famous quote about the relevance of animal suffering given in Chapter 2, was a mere footnote, and not a comprehensive theory. Indeed, a fully worked out utilitarian treatise promoting justice for animals did not really arrive until the work of Peter Singer in the 1970s. As such, the main focus of this chapter will be the work of Singer and his critics. The first section of the chapter provides an overview of Singer's theory, and its implications for policy issues such as animal agriculture and animal research. The chapter then goes on to examine a number of critiques of Singer's utilitarian theory. First of all, it will examine critiques which say that Singer goes too far and grants too much weight to the interests of animals. Secondly, it will examine critiques that claim that Singer's policy proposals do not logically follow from his theory. Finally, the chapter will examine claims that Singer's theory offers insufficient protection to animals, and what is really required is a theory of animal *rights*. First of all then, let us examine Singer's famous and influential theory of animal liberation in more detail.

Singer's utilitarian theory of animal liberation

We have seen that concern about cruelty to animals, and concern about our political obligations to animals, stretches back to Ancient times. However, it is also clear that such concern has come in waves. One important wave came in the eighteenth and nineteenth centuries, as many European philosophers, campaigners and legislators sought to outlaw practices involving animals that were thought to be cruel. This wave was undoubtedly influenced by the suffering of animals that could be seen on the streets of the time: exhausted horses pulling loads beyond their capacity, or cattle being brutally driven to the market and slaughterhouses. Another important wave came in the middle of the twentieth century. This time, the wave was primarily stimulated by growing concern about modern farming practices, and their impact on animal welfare. After all, animal agriculture went through something of a revolution in the post-war years. As the population of industrialised countries expanded, as consumer wealth grew and as new technological developments emerged, efforts were made to rapidly increase the efficiency of meat production. These efforts to make meat production more efficient led to the rise of 'intensive' farming methods, and the development of the so-called 'factory farms'. Such farms were far removed from the image of open pasture and pastoral care that were and continue to be imprinted in the minds of many consumers. Instead, they were industrial processing plants, with animals confined in sheds, cages, stalls and crates, and subjected to all sorts of painful practices over the course of their short lives. Animals had become production units. As a result, consumers enjoyed a more plentiful supply of cheaper meat; while animals endured short and torrid lives. It was Ruth Harrison's book, *Animal Machines,* published in 1964 which first opened the public's eyes to the appalling realities of these factory farms.[1] That book not only stimulated public concern about such farms, but it also helped to create a renewed interest amongst political theorists about the question of what political communities owe to animals. Indeed, it was read by the utilitarian philosopher Peter Singer and influenced his own important 1975 publication, *Animal Liberation,* which has been deemed as 'the bible of the animal rights movement'.[2]

Singer's theory is interesting in the way that it both uses and changes the classical utilitarian theory of Bentham. Singer, like Bentham, argues that the relevant characteristic for being included in considerations of morality and justice is sentience: the capacity to feel pleasure and pain. As we saw in the previous chapter, Bentham argued that the morally relevant characteristic cannot be something like reason, free will or language use. After all, young infants lack such capacities, and yet we do include them in our considerations when formulating moral and political obligations. We rightly include them, Bentham claimed, because they can *suffer*: it is this capacity that is of moral relevance, and it is this capacity which makes one deserving of justice according to Bentham. While Bentham left his discussion of animals pretty much at that, Singer builds on it to develop a principle which he calls 'the equal consideration of interests'. According to Singer, if an entity is sentient, then that entity should be given equal consideration when we formulate our obligations. For Singer, every interest of every sentient being must be considered on its own merits. To prioritise the interests of a particular race would be racist, to prioritise the interests of a particular gender would be sexist, and to prioritise the interests of a particular species would be *speciesist*. Of course, it is important to note that Singer's claim is not that every sentient being should be *treated* equally; but rather, that every sentient being should be *considered* equally. Thus, for Singer it would not be wrong to deny pigs the vote, for obviously pigs have no interest in participating in a democratic society; but it would be wrong to disregard pigs' interest in not suffering, for clearly pigs have a strong interest in avoiding pain, just like us. In order to establish what obligations and policies follow from this principle of equal consideration, Singer combines it with the utilitarian goal of promoting utility, or in his terms, satisfaction of interests. As such, there are two strands to Singer's theory: first of all, we must consider the interests of sentient beings equally; and secondly, our actions and policies should aim to bring about the greatest amount of interest-satisfaction possible.

What policy prescriptions follow from Singer's utilitarian theory? The two issues that Singer concentrates on in *Animal Liberation* are animal experiments and meat consumption. Let us start with the issue of eating meat. Singer asks us to consider the interests of all sentient beings equally. When it comes to the issue of eating meat, two rival interests are plainly evident: the human interest in enjoying eating meat; and the animal interest in avoiding the sufferings of the practices of the factory farm. What does the principle of equal consideration of interests say about them? First of all, it mandates that we cannot simply grant priority to the human interest in eating meat. It states that ignoring or relegating the interests of animals simply because they belong to non-humans is speciesist and impermissible. Instead, the principle obliges us to consider interests on their own merits, and to give equal weight to equal interests, irrespective of who they belong to. Importantly, Singer argues that the interest in eating meat and the interest in not suffering are not of equal weight. Singer claims that the human interest in eating meat is only trivial on the bases that humans can lead perfectly healthy lives without eating meat and can gain plenty of pleasures of the palate from non-meat alternatives. The animal interest in not suffering, on the other hand, is major. The pains inflicted by factory farms are intense and real, and all sentient beings have a strong interest in avoiding pains of these kind. Singer then claims that the principle of equal consideration of interests does not allow for major interests to be sacrificed for trivial interests, and thus that maximum interest-satisfaction would be achieved by the abolition of factory farms. Furthermore, Singer argues that this abolition can be achieved if we all stop eating the food that factory farms produce.[3]

When we turn to the issue of animal experimentation, however, we find that the relevant competing interests are more complex. First of all, Singer points out that some animal experiments pitch trivial human interests against major animal ones, as was the case with eating meat. This is the situation, Singer claims, with painful non-therapeutic experiments that test the safety of cosmetics, detergents or food additives, or which produce fairly useless knowledge about animal behaviour. As such, according to Singer, the principle of equal consideration of interests demands that all such experiments come to an end.[4] Such experiments use animals simply as a resource, fail to give due weight to their interest in not suffering and fail to promote utility under Singer's theory. However, the human interests in the benefits afforded by painful *therapeutic* animal experiments are obviously not so trivial.[5] Human beings have strong interests in avoiding illness, in being cured of illness, and in being relieved of the pain and suffering of illness. Animal experiments which research diseases and which test drugs to cure or mitigate the effects of diseases thus seem to satisfy major and important interests of human beings. What does Singer's theory have to say about such therapeutic experiments?

In the first place, it is worth remembering that under Singer's theory the suffering of the animals involved in therapeutic experiments counts, and counts equally to the suffering that human beings endure from illnesses. In which case, in order to justify painful experimentation on animals, the onus is on the experimenters to show that their actions will alleviate more suffering for both humans and animals in the long term. This would involve a radical change in policy for most states, and lead to far fewer painful animal experiments across the globe. After all, this burden of proof is clearly too great for many current animal experiments, and certainly rules out those that are purely speculative and which only have a slim chance of providing anything useful. Secondly, it is nevertheless clear that Singer's theory does not rule out animal experimentation as a matter of principle. It is quite possible that interest-satisfaction could be maximised by a painful experiment on a group of animals. As Singer himself states:

> ... if one, or even a dozen animals had to suffer experiments in order to save thousands, I would think it right and in accordance with equal consideration of interests that they should do so. This, at any rate, is the answer a utilitarian must give.[6]

However, states should not take this to mean that those animal experiments which can show that the benefits will outweigh the costs are morally unproblematic. After all, why is it that animals and not human beings must suffer for the sake of promoting overall utility? One cannot simply appeal to the fact that human beings are more rational, or belong to families that care deeply for them, for, as Singer points out, these facts are not true for all humans. Accordingly, Singer claims:

> If experimenters are not prepared to use orphaned humans with severe and irreversible brain damage, their readiness to use nonhuman animals seems to discriminate on the basis of species alone....[7]

For Singer then, if one is prepared to justify painful experimentation on a few animals on the basis that it promotes utility for the many, one must be ready to apply that same principle to painful experimentation on human beings. To fail to do so would be to prioritise human interests without good reason. Such prioritisation is arbitrary, speciesist and is ruled out under Singer's theory.

Conservative critics of Singer's theory

Most readers new to Singer's theory will find it controversial. The principle of equal consideration of interests is radical. It is in direct opposition to the traditional social norms of many contemporary societies, which state that human beings and their interests are simply more important than non-human animals and their interests. Its policy proposals are also radical. They demand the end of intensive animal agriculture, and the end of the vast majority of animal experiments. As such, it will come as little surprise to learn that Singer's theory has been subjected to a number of critiques. While there is insufficient room to discuss all those critiques here, it will be useful to examine some of the most important. To this end, this section looks at conservative critics of Singer's theory, who say his theory affords too much protection to animals. The following section looks at critics who claim that Singer's policy proposals do not follow from his theory. And the final section looks at radical critics, who say that his theory offers insufficient protection to animals.

One important criticism of Singer's theory denies that animals possess interests. If correct, this criticism would be devastating to Singer's theory, premised as it is on the equal consideration of interests and the goal of maximising interest-satisfaction. If animals do not count on the register of interests, then there is no need to consider them at all when formulating policies aimed at maximising interest-satisfaction: we could treat them as we liked, in farms, laboratories and elsewhere. While it does sound odd to deny that animals possess interests, important arguments to this effect have been provided by the philosophers McCloskey and Frey. Both McCloskey and Frey argue that for an entity to have interests means more than certain events are good or bad for it. Of course, certain events are good or bad for animals, like having enough to eat and being caused pain. But as they point out, some events are also good or bad for inanimate objects like plants and machines. Watering a plant seems to be good for a plant; just as oiling a machine is good for a machine. And yet, most of us do not think that plants and machines possess morally relevant interests. For McCloskey then, to have an interest in some good, one not merely has to be benefited by the realisation of that good, one also has to be *concerned* about that good.[8] For Frey, to have an interest in some good, one not merely has to be benefited by the realisation of that good, but one also has to *desire* that good.[9] To have interests is to have some active concern and desire for certain goods, something which both authors claim that animals lack.

However, even if we accept McCloskey and Frey's definition of interests, it still seems that sentient animals possess them. After all, everyday observation seems to confirm that sentient animals *do* desire and *are* concerned about what is good for them. Moreover, such observation also suggests that sentient animals are concerned about and desire such goods in ways that plants and machines do not. Food, shelter, love, play, exercise and so on do not merely benefit a dog like oil benefits a tractor. For not only do dogs *feel* the benefits of such goods, but they also behave as if they desire and have concern for those goods. Consider, for example, a dog begging for food. Surely that dog is concerned about eating and desires to eat. Of course, without actually entering the mind of the dog ourselves, it is impossible to be *absolutely* certain of this. There is of course a chance that the dog is behaving mechanistically, and feels absolutely nothing. But there is the same chance that all other human beings feel nothing and behave purely mechanistically. And yet, most of us do not think that all other humans are unfeeling robots. We infer from their behaviour and what we know about their anatomy that they consciously desire and are concerned about what is good for them. And if we do this for human beings, surely it is only reasonable to do the same for animals: that is, we can infer from their behaviour and what we know about their anatomy that they consciously desire and are concerned about certain goods. In light of all this, if it is necessary to desire and be concerned about a good in order to have an interest in that good, then it seems that animals do possess interests, as Singer assumes.

The second important criticism of Singer's theory is that the principle of equal consideration of interests is false. Some theorists have argued that it is perfectly legitimate for human beings to prioritise the interests of their fellow beings. As such, they claim that speciesism, far from being on a par with racism or sexism, is actually quite permissible. One recent attempt to justify speciesism comes from Lewis Petrinovich. Petrinovich argues that favouring one's own species is a natural fact of life:

> Humans, as well as all other social animals, are speciesists. Animals of all species show a clear preference for their own kind: They associate and mate with their own species; they fight alongside their own kind against members of a foreign species to secure resources; and they defend the young of their own species. Any species that did not show preference for its own kind would become extinct.[10]

Furthermore, Petrinovich argues that because of this alleged fact, '... the interests of members of our species should triumph over comparable interests of members of other species'.[11]

However, there are some important problems with this type of argument. In the first place, one can question the dubious picture of species solidarity that Petrinovich paints. After all, individual animals do associate and mate with individuals from other species, they do fight and kill members of their own species, and they even eat the young of their own species. As such, favouring one's own species in not a clear and unambiguous natural fact of life as Petrinovich suggests. Secondly, even if it were, what does that prove ethically? After all, just because something is a natural fact does not mean that it is therefore ethically right. Rape and murder are surely natural facts of life, but quite obviously that does not make them morally permissible. Finally, we must ask what makes favouring one's own *species* permissible. After all, we all belong to a wide variety of biological and social groups, and many of us have stronger bonds with these groups than we do with our species. And yet, few claim that it is permissible for whites to favour the interests of their fellow race, or women to favour the interests of their fellow gender. It thus seems odd to claim that favouring one's own species is permissible, but that favouring one's own race, gender, class and so on is impermissible. As such, it would appear that Singer is right to suggest that speciesism is morally problematic, because of the simple fact that species is not a morally relevant category.

Critics of Singer's policy proposals

Another line of criticism of Singer's theory does not contest the fact that animals possess interests, nor does it contest the principle of equal consideration of interests. What it does contest is the policy proposals that flow from them. This objection is important because it reveals and relates to the *consequentialist* nature of utilitarianism. A consequentialist theory assesses actions and policies solely by their consequences and not, say, by their conformity to some rule or ordinance. Utilitarianism, as we have seen, judges policies by their impact on overall utility or interest-satisfaction. Utilitarians claim that this gives their theory a certain plausibility, grounding it in empirical facts about the real world, as opposed to say abstract and mysterious principles. However, the consequences of actions and policies, and their affect on utility, can always be contested. And as we shall see, Singer's claims about the consequences of abolishing the meat industry have been contested. Would the abolition of the meat industry really promote utility?

Some philosophers have questioned whether meat-eating really does pit trivial human interests against major animal interests, as Singer claims.[12] For example, a number of thinkers have questioned whether the human interest in eating meat is really so trivial. After all, widespread resistance to adopting vegetarianism, even amongst those who are perfectly aware of the suffering that intensive farming practices cause, might be cited as evidence of the depth and strength of the human interest in eating meat. Moreover, eating meat is not the only human interest at stake. The meat industry is big business. If that industry were to close down, there would be obvious deleterious effects to the economy and employment. Perhaps then, when all the relevant interests are taken into account and given their proper weight, the costs of closing down the meat industry might well in fact outweigh the benefits. It might be then that Singer's utilitarian theory actually supports the continuation of the meat industry, rather than its abolition.

However, such objections are not very impressive. When Singer describes the human interest in eating meat as trivial, he does not mean to deny that many humans take intense pleasure from eating meat. All he means to convey is that human beings can ordinarily lead lives of extremely high quality without eating meat. Animals, in contrast, cannot lead lives of much quality at all when they are subjected to the kinds of practices and confinement of intensive farming. Furthermore, while it is inevitably true that many human beings have an interest in the continuation of the meat industry for economic reasons, one has to remember that the animal interest in not suffering counts equally. And when one considers that in North America and Europe alone, something like *17 billion* land animals are raised and killed for food annually, it seems plausible that the benefits of closing down the meat industry outweigh the costs when all sentient beings are given equal consideration.[13]

But perhaps this is too hasty. After all, as several philosophers have pointed out, if there had been no meat industry then those 17 billion animals would never have existed at all. For, quite obviously, those animals were created by the industry solely for their meat. In this way, the industry actually creates more lives in the world, and thus more utility. Yes, it is true that the animals raised by the industry are killed, but those animals are continually 'replaced' by new animals. Perhaps then, and as some philosophers have claimed, a world with a meat industry has greater overall utility than a world without a meat industry. As such, perhaps the meat industry should actually be supported under Singer's utilitarian theory.

Of course, in order for this argument to work, it must be shown that those animals which the meat industry brings into existence actually gain some benefit from their lives. For if the animals endure lives of such misery that their lives are not worth living, then their creation actually creates disutility, and thus should not be permitted under a utilitarian theory. Indeed, given the terrible suffering that many animals endure on factory farms, it is perfectly possible to argue that many animals raised for their meat do lead lives that are not worth living. However, let us for the sake of argument assume that it is possible to raise farm animals with lives that are at least minimally worth living, such as those living under free-range conditions. Would Singer's theory actually support the raising and killing of free-range animals for their meat?

Singer argues that it would not. And he does so by emphasising that he is not a classical utilitarian, but a preference utilitarian. Thus, unlike thinkers such as Bentham, Singer believes that utility cannot simply be defined in terms of levels of pleasure and pain, but also in terms of satisfaction of preferences.[14] Thus, even if it were true that a world with a meat industry had more pleasure overall than a world without it, that does not decide the issue for Singer. For Singer is not primarily concerned with whether the meat industry produces more experiences of pleasure, but how it fares in bringing about what is preferred.

When evaluated on this basis, Singer claims that the meat industry does rather badly. To explain, Singer draws a distinction between 'self-conscious' entities and 'merely conscious' entities. Singer believes that self-conscious entities are those beings who are not merely sentient, but who are also aware of themselves existing over time. In other words, they are beings who can possess memories of their own past, and who can forge plans and desires about their future lives. Singer claims that self-conscious animals do not only have interests in avoiding pain and pursuing pleasure, but also have an interest in pursuing and realising their future-oriented desires, including the desire to continue to live. Singer then claims that when a self-conscious animal is killed, all of its future-oriented desires are thwarted. Of course, it is possible to create a new animal with new opportunities for pleasure and new desires. However, that does not compensate for the loss in utility that the killing caused: the original animal's desires remain unsatisfied. The creation and slaughter of self-conscious animals creates disutility by thwarting huge numbers of desires, and is wrong as such according to Singer.

And yet, even if we buy into this argument, it obviously only works for 'self-conscious' animals. Singer himself admits that merely conscious animals have no capacity to conceive of themselves as existing over time, and thus have no future-oriented desires. The killing of these animals then, Singer claims, thwarts no desires, and would only be problematic in so far as it decreases the overall experience of pleasure in the world. As such, so long as merely conscious animals lead worthwhile lives, and are replaced by new animals, then there is nothing wrong in itself with killing them, according to Singer's theory. Given this surprising turn, it obviously becomes rather important to determine just which animals are self-conscious and which are merely conscious. Singer claims that a case can be made for thinking that all mammals are self-conscious beings, thus making the raising and killing of such animals for their meat impermissible. However, he believes that no such case can be made for birds and fish. Assuming then that Singer is correct on this highly contentious point, we see that his theory can actually support the raising of fish and poultry for their meat, so long as they lead lives that are worth living, so long as they are killed painlessly, and so long as they are replaced by animals who would lead equally pleasant lives.

But while Singer accepts this conclusion in *theory*, in *practice* he argues that it is better to reject *all* killing of animals for food. This is based on a 'slippery slope' type of argument. He argues as follows:

'To foster the right attitudes of consideration for animals, including non-self-conscious ones, it may be best to make it a simple principle to avoid killing them for food.'[15] To explain, Singer believes that if we abolished the killing of mammals, and were left with farms that raised fish and poultry in free-range conditions, we would come to regard those fish and poultry as mere resources, which would in turn cause us to slide back into intensive farming methods, perhaps even for mammals too. This claim, however, is highly questionable. After all, if a political community were to adopt an agricultural system in which the only animals farmed for food were free-range fish and poultry, a radical shift in attitudes surrounding animal welfare would have had to have taken place. The individuals of that society would have had to have altered their valuation of mammals in a profound way. I wish to contend that it is extremely unlikely that a political community which had undergone such a transformation would slip back into the kind of intensive farming methods in the way that Singer suggests.

Whatever the truth regarding the potential for sliding back to intensive farming practices, a wider point about Singer's theory and utilitarianism as a whole should be evident. Just how are political communities to accurately assess the consequences of actions and policies? As the above discussion shows, the question of whether abolishing the meat industry would create more utility in the world is far from straightforward. The consequences of different policies are often hard to ascertain, and can often be surprising. Indeed, several philosophers have recently questioned whether a switch to universal vegetarianism would even result in fewer animals being killed. After all, animals are still killed in the production of crops. This is quite simply because many field animals – birds, mice and other small rodents for example – are killed through ploughing the fields and cultivating the crops. Because of this fact, Steven Davis and George Schedler have recently argued that the agricultural system which in fact kills the fewest animals would not be a vegetarian crop-only one, but an omnivorous one, involving both crops and ruminant pasture. They point out that keeping large herbivores such as cattle on fields reduces the number of times that the fields need to be worked. As such, they claim that an agricultural system which reserves some land for ruminant pasture would actually kill the fewest animals.[16] Inevitably, however, another utilitarian has denied these claims and argued that these authors get their sums wrong. Gaverick Matheny argues that because crops require so much less land than cattle in order to produce the same amount of protein, the agricultural system which kills the fewest animals would still be a crop-only one.[17]

Whatever the truth, it is clear that the policy implications of utilitarian theories of justice require careful assessment of the future consequences. While it is undoubtedly the case that many of those consequences are somewhat ambiguous, it is interesting to note where consensus lies. Different thinkers propose different agricultural models. But whether they propose a system which kills no animals at all, whether it be one with free-range fish and poultry, or whether it be one with some ruminant pasture, all agree that utility is not promoted by the *existing* system. All agree that intensive farming methods create disutility and should be abolished as a result.

Radical critics of Singer's theory

The final set of criticisms that we need to address claim that Singer's theory, and all other forms of utilitarianism, far from being radical in their extension of justice to animals, in fact fail to grant animals adequate protection. In this section, we will consider two important types of criticism along these lines: the first claims that utilitarianism neglects certain harmful ways of treating animals; and the second states that utilitarianism fails to impose any absolute limits on the treatment of animals.

The first criticism claims that the utilitarian protection of animals is insufficient because it neglects certain harmful ways of treating animals. This critique claims that there is more to animal well-being than simply how they feel. It argues that utilitarianism mistakenly judges animals' lives solely by their conscious states: whether they are suffering or whether they have their preferences satisfied. However, some thinkers have claimed that animals can suffer from unfelt harms, and benefit from unfelt goods. For example, one recent proponent of this view is Martha C. Nussbaum:

> It seems plausible to think that there may be goods they (animals) pursue that are not felt as pain and frustration when they are absent: for example, free movement and physical achievement, and also altruistic sacrifice for kin and group. It is also possible that some animal pains may even be valuable: the grief of an animal for a dead child or parent, or for the suffering of a human friend, may be a constituent part of an attachment that is intrinsically good, as may the pain involved in the effort required to master a difficult activity.[18]

Nussbaum claims that what is crucial for animals' well-being is that they possess the opportunity to exercise the natural 'capabilities' of their species. Having the ability to exercise such capabilities, she argues, is necessary for an animal to flourish as the kind of being that it is and lead a truly dignified existence. As such, Nussbaum claims that the goal of political communities in relation to animals should be to formulate policies which provide them the opportunity to lead such dignified lives. Obviously, minimising pain and promoting preference-satisfaction might be part of such policies. But for Nussbaum, the achievement of such conscious states does not exhaust our policy goals: animals should be granted the opportunity to exercise their natural capabilities, allowing them to flourish as members of their own species.

To explain how Nussbaum's 'capabilities approach' differs from utilitarianism, consider the practice of dressing bears in human clothes and getting them to perform tricks, as carried out by certain circuses.[19] A purely utilitarian assessment of such practices considers only how those circus acts affect overall utility. Does the training of the bears, the way that they are kept, their adornment with human clothes or their performance to audiences cause the bears to suffer in any way? If not, and other things being equal, then a utilitarian would ordinarily consider such practices to be permissible. The capabilities approach, however, does not just weigh up conscious states in this way. The capabilities approach also asks whether these bears are leading dignified lives and flourishing as the types of being that they are. And given that these bears are held captive, dressed as humans and made to perform tricks, it is certainly possible to argue that they are not flourishing members of their species. It may be the case, if somewhat unlikely, that these bears do not suffer. Nevertheless, the capabilities approach sees something wasteful and tragic in the simple fact that these bears are not flourishing *as bears*: they cannot forage, hunt, mate and so on as flourishing members of their species ought to do. As such, by focusing on the opportunities that animals possess to exercise their natural functionings, the capabilities approach claims to offer greater protection to animals than utilitarian theories.

However, the capabilities approach is not without its own difficulties. In the first place, the move from conscious states to natural capabilities has more radical implications than Nussbaum herself is prepared to admit. After all, the shift away from a sole focus on sentience leads not only to an expansion in the types of *harm* under consideration, but also the types of *entity* under consideration. Nussbaum claims that we should formulate policies which enable entities to lead flourishing lives as members of their own species, because there is, '...something wonderful and wonder-inspiring in all the complex forms of life in nature'.[20] But of course, if it is the flourishing of natural life that is crucial to considerations of justice, this means that we should extend justice to all living organisms, and not merely to sentient animals. Some radical environmentalists would be happy to accept this conclusion.[21] And yet, is the flourishing of bacteria, viruses, blades of grass, mosquitoes and so on something which political communities should really strive to promote? Indeed, Nussbaum herself thinks not, and thus limits her capabilities approach so that it applies only to sentient life: 'I believe that we have enough on our plate if we focus for the time being on sentient creatures.'[22] It seems then that Nussbaum draws the line at sentience simply for purely practical purposes. She would be happy to extend justice to all forms of living organisms, if it were feasible. But this does open up the question of whether Nussbaum is fetishising nature in her theory of justice. After all, it is surely extremely doubtful that all functionings and capabilities of all species are valuable and worthy of promotion, even if such promotion were feasible. Most societies quite rightly refrain from promoting the murderous capacities of human beings, the spread of disease-carrying parasites and the flourishing of harmful viruses. It seems obvious that some natural functionings should be inhibited rather than fostered. In fact, Nussbaum admits this much herself and claims that the capabilities approach is not 'nature worship'. Instead, she argues that we must discriminate between capabilities, and judge only some as worthy of being fostered.[23] But this creates difficulties for her theory. For once we move away from the idea that the natural functionings of entities are valuable in and of themselves, the method for judging just which capabilities are valuable and which are not is somewhat mysterious. Peter Singer has even suggested that one obvious way to judge which capabilities are valuable and which are not would be to assess how they affect welfare. This of course would make the capabilities approach just another type of utilitarianism![24] Following this line of thought then, it might be argued that the only way that the capabilties approach can be made plausible and straightforward is to revert back to some form of utilitarianism.

The second radical criticism, however, would refute strongly the idea that in the end, alternative theories of harm must fall back on utilitarianism. This critique claims that the utilitarian protection of animals is insufficient because that protection can only ever be *contingent*. Utilitarians recommend policies which prevent animals from suffering or from being killed *only if* such policies promote welfare. For a utilitarian, there are no absolute limits on what we can do to animals; no actions are categorically ruled out in principle. For some proponents of justice for animals, this situation leaves animals extremely vulnerable to a range of harms. The most famous proponent of this view is Tom Regan. Regan points out that for utilitarians, the ultimate object of value is welfare, rather than the individuals who experience that welfare. As such, sentient animals are valued only in so far as they are 'receptacles' of welfare. This way of viewing animals is clearly evident in Singer's writings about merely conscious animals like birds and fish, who he admits are 'replaceable', even if only in theory. Regan claims that this way of valuing individual animals is completely wrong. Regan argues that those animals who are 'subjects-of-a-life' – those entities with beliefs and desires, perception and memory, the ability to feel pain and pleasure, the ability to initiate action and so on – possess 'inherent value': a value which is not reducible to their contribution to utility. As such, Regan claims that all entities with inherent value have a right to respectful treatment: 'We are to treat those individuals who have inherent value in ways that respect their inherent value.'[25] Essentially, this is a Kantian injunction whereby, '. . . individuals who have inherent value must never be treated *merely as means* to securing the best aggregate consequences'.[26]

What Regan is putting forward here is radical indeed. Singer essentially argues that animals should be included in the cost-benefit utility calculations of individuals and communities. Regan argues that this does not go far enough. Instead, he argues that because of something fundamental about their nature, animals have rights which cannot be transgressed in the name of overall utility. Thus, even if it turned out to be the case that animal experiments, the meat industry, performing bears and so on actually increased overall utility, for Regan, such practices would still be impermissible. Such practices are impermissible, he claims, because they fail to respect the inherent value of individual animals, and because they treat animals merely as means to achieving some set of desired consequences. Regan argues that just as there are some things like torture, cannibalism, rape and so on, which should never be done to human beings, whatever the benefits that might accrue, so there are some things which should never be done to animals. As such, Regan argues that political communities should not only include animals in their deliberations about justice, but should protect this basic animal right to respectful treatment.

However, this attribution of rights to animals is extremely controversial. Theorists such as Michael Fox and Carl Cohen have claimed that rights are necessarily *reciprocal,* and thus cannot be attributed to animals. They argue that because rights entail correlative duties, the possession of rights requires the ability to fulfil those correlative duties.[27] On this logic, it cannot be true that a chicken has a right that I do not kill it to eat it, for we do not stand in a reciprocal relationship. The chicken can neither understand nor respect my right not to be killed; it can hold no duties towards me; and is thus argued to hold no right against me.

The problem with this reciprocal understanding of rights, however, is that it seems to lead to the exclusion of many human beings from the possession of rights. After all, and as Regan himself points out, young infants and the severely mentally disabled are not able to understand or respect rights, yet most of us believe that they are the sort of entities who can possess them. In response to such arguments, Fox and Cohen claim that we should ignore such exceptional cases. The important point for them is that such human beings belong to a wider group, the majority of whom *do* have the necessary capacities for the possession of rights. As such, they claim that all human beings should benefit from being awarded rights.[28] But the great problem with this argument is that individuals belong to all sorts of wider groups, and it is unclear why species should be considered the one that is decisive. Indeed, Nathan Nobis has pointed out that the logic of Fox and Cohen can be used to deny rights to *humans.* To explain, consider the simple fact that human beings are living organisms. They thus belong to a wider group who do not ordinarily possess the capacity to reciprocate. If we assign rights not on the basis of the capacities of individuals themselves, but the general capacities of their wider group, then if that wider group is 'living organisms', it seems that we should deny rights to human beings.[29] As Nobis correctly states, there are grave problems with assigning rights on the basis of the characteristics generally held by wider groups, and not on those of individuals themselves.

It would seem then that a means of assigning rights that relies on the characteristics of individuals is more persuasive. And indeed, under Regan's theory, all individuals who are 'subjects-of-a-life' possess rights. To explain in more detail, Regan gives the following definition of subjects-of-a-life:

... individuals are subjects-of-a-life if they have beliefs and desires; perception, memory and a sense of future, including their own future; an emotional life together with feelings of pleasure and pain; preference and welfare interests; the ability to initiate action in pursuit of their desires and goals; a psychophysical identity over time;

and an individual welfare in the sense that their experiential life fares well or ill for them, logically independently of their utility for others and logically independently of their being the object of anyone else's interests.[30]

So which individuals satisfy Regan's 'subject-of-a-life' criteria? Such line-drawing is notoriously difficult, but nevertheless important in practice. Regan thinks that we can be confident that mammals above the age of one are subjects-of-a-life.[31] While he does state that this does not mean that *only* mammals above the age of one possess inherent value, it is clear that while Regan's theory of rights is more inclusive than one based on reciprocity, its logic means that he will necessarily have to exclude some sentient animals, and indeed some sentient humans.[32] This will be troubling for those proponents of animal rights, and indeed human rights, who believe that all sentient individuals merit the protection that rights afford.

One way of dealing with this problem is to avoid grounding rights in the inherent value of subjects-of-a-life, but instead to ground them in 'interests'. That is to say, it is possible to argue that some aspects of sentient individuals' well-being are so important that they should be protected in the form of rights. Indeed, this proposal was made in an early article advocating animal rights by Joel Feinberg.[33] This interest-based rights approach has a number of advantages, and can even be seen as a compromise between utilitarianism and rights-based theories. In the first place, it mirrors utilitarianism in making sentience itself the simple and tangible basis for inclusion. Under an interest-based theory of rights, all entities with interests are potential bearers of rights, and for most philosophers that means that all sentient beings are potential bearers of rights. Secondly, because it is a rights theory, it does not simply seek to maximise the interest-satisfaction of society. Instead, it recognises that there are some limits to this utility maximisation: some interests of individuals are so important that they cannot be sacrificed for greater social utility. As such, interest-based rights provide means of combining the wisdom of utilitarianism in terms of who merits justice with the wisdom of rights theory in terms of placing some limits on untrammelled utility maximisation. Of course, none of this suggests that an interest-based rights approach is without problems of its own. For one, any such theory needs to provide a mechanism for determining just which interests are important enough to be granted protection in the form of rights. For example, it is relatively easy to acknowledge that sentient animals have a strong interest in not suffering, but does that mean that all sentient animals have a right not to suffer in all contexts? Does a gazelle being chased by a lion have a right not to be made to suffer in the same way that a pig in a factory farm has a right not to be made to suffer? Clearly, any theory of interest-based rights will need a means of resolving such difficult questions.[34]

As a final point concerning the issue of rights, it is also worth considering whether utilitarianism must necessarily be opposed to the idea of granting legal rights to animals in political communities. After all, utilitarianism seeks policies that promote utility. Rights-theorists like Regan claim that this leaves individuals vulnerable. For example, if the sacrifice of a few individuals in painful experiments cures thousands of a painful disease, then utilitarianism must endorse such experimentation. However, we have seen that assessing the consequences of policies is a complex business. For one, it is important to consider that there may be differences between short-term and long-term assessments of utility. To explain, it is worth noting that many utilitarians argue that human beings should be granted an absolute legal right not be tortured. For while it may be possible to imagine a set of circumstances in which the torture of an individual could lead to better overall consequences in the short term – the classic example of torturing a terrorist to find the location of a ticking time-bomb is obvious – the long-term effects of making torture legally permissible are likely to be disastrous. For not only does torture rarely supply reliable information; not only are legal procedures for torture likely to be stretched and abused; but it has also been argued that legalising torture would completely erode citizens' faith in the justice and fairness of their basic political institutions.[35] Using similar reasoning, it may be possible for utilitarians to endorse policies which protect a set of animal rights. For even if it were true that animal experiments, the meat industry or performing bears promoted immediate utility, perhaps their long-term effects actually create more costs. As Singer himself states with regard to the meat industry, it could well be that such practices help sustain the view that animals are mere resources that we can use as we please, encourage other and further cruelties to animals, and thus lead to an overall reduction in utility in the long term. On just such a basis a utilitarian could well support a set of legal rights for animals.

Conclusion

In this chapter, we have seen how utilitarianism offers a simple means of including animals in deliberations about justice. Because utilitarianism is welfarist and egalitarian, it can readily include the interests of sentient animals in its deliberations about which policies best promote utility. The most comprehensive utilitarian theory that has attempted to extend justice to animals is that of Peter Singer. Singer's theory revolutionised attitudes to animals. It offers a direct challenge to other political and moral theories which unthinkingly prioritise the interests of human beings, and it offers a direct challenge to the policies of states with its calls to abolish the meat industry and most forms of animal experimentation. But these radical proposals have inevitably met with a great deal of critical response. Some have argued that Singer grants too much protection to animals: they either claim that animals lack morally relevant interests, or that those interests can legitimately be subordinated to those of humans. Others have argued that Singer gets his policy proposals wrong: they claim that the total abolition of the meat industry would not necessarily promote utility. Finally, still others have claimed that no utilitarian theory can grant sufficient protection to animals: they either claim that utilitarianism neglects certain important harms, or they claim that it fails to sanction any absolute prohibitions on the use of animals. All of these objections are important, and many have taken them to be decisive, dismissing utilitarian political theories out of hand. But utilitarianism is a hardy theory and continues to appeal to many theorists. After all, it does offer a clear and tangible basis with which to formulate policies. Moreover, more sophisticated forms of utilitarianism are available than its classical form: definitions of utility can be subtle and assessments of consequences can be long term. But despite this hardiness, it nevertheless remains the case that utilitarianism fell from its dominant position in Anglo-American political theory just around the time that Singer was writing. For many theorists, its sole focus on overall welfare, rather than individuals, was too problematic. As such, from the middle of the twentieth century, liberalism came to dominate political theory. And as we will see in the next chapter, this had a major impact on the treatment of animals in the discipline.

4
Liberalism and Animals

In the previous two chapters we saw that the rise of utilitarianism led to an increased concern for animals amongst political theorists. While writers such as Hutcheson, Primatt and Bentham all made the link between sentience and justice, the first comprehensive utilitarian theory of animal justice did not arrive until the publication of Peter Singer's *Animal Liberation* in the 1970s. However, we also saw how one group of critics regarded utilitarianism to be flawed as a basis to protect animals. These critics argued that utilitarianism was flawed because of its focus on aggregate utility and its failure to offer any absolute prohibitions on the treatment of individual animals. Critics like Tom Regan claimed that we need a theory which recognises the inherent value and rights of individual animals.

But this concern to provide a more robust protection for individuals than that provided by utilitarianism is not unique to debates about animals. In fact, it is a familiar position in political theory more generally. Moreover, it is a position made famous by John Rawls in his work, *A Theory of Justice*, first published in 1971. In that work Rawls was primarily concerned to offer a defence of liberal democratic institutions. However, he was keen to provide that defence without basing it on some form of utilitarianism. After all, in the first half of the twentieth century, utilitarianism was the dominant school in Anglo-American political theory. This domination reflected a rise in the scientific study of politics, and hence a favouring of disciplines such as economics and social science over normative political theory. As such, around this time it was thought that devising the appropriate policies for political communities was best achieved not through abstract theorising, but through some form of technical calculation, like the utilitarian method of cost-benefit analysis. However, at the height of the Cold War, Rawls and other liberal theorists came to believe that it was particularly important to provide a justification for liberal democratic institutions that was not contingent on their promotion of overall welfare. For John Rawls in particular, liberalism needed a more robust defence than utilitarianism could offer. Utilitarianism offered an insufficient justification for liberal institutions for Rawls, because it failed to take seriously the 'separateness of persons'. In other words, and as we saw in the last chapter, utilitarianism can justify policies which harm individuals if those actions contribute to overall welfare. As such, in *A Theory of Justice* Rawls sought to provide and defend political principles which took the value of 'persons' more seriously.

I want to argue that this valuation of individual persons is the defining feature of liberalism. Unlike utilitarians, who place ultimate value in aggregative welfare, the fundamental object of value for liberals of all stripes is the free and equal individual person. Under liberal theory, individuals are considered as free on the basis that they are autonomous agents who can reason about, reflect on and pursue their own goals in life. In the terminology of liberals, human beings are 'persons', and it is their personhood which makes them distinctive, and provides the reason for holding humans in such esteem. So, while modern utilitarians have as their key influence the work of Jeremy Bentham, many modern liberals, including John Rawls, have as their key influence the work of Immanuel Kant. For you will recall from Chapter 2 how Kant claimed that human beings have dignity because they are 'ends-in-themselves': that is, they are persons who have the ability to reason, act morally and frame and pursue their own life goals.

Moreover, the free individual person is also considered to be equal under liberal theory. Liberals are not claiming that individuals are equal in terms of their characteristics and capacities; the claim of equality is not an empirical claim. Instead, the claim is that persons are *moral* equals, with none possessing any inherent moral superiority, and none inherently born to rule while others serve. This egalitarianism is clearly testament to liberalism's historical roots. After all, liberalism emerged as a distinctive political theory just as seventeenth-century European communities were beginning to question the traditional hierarchies of medieval society. The notion that all persons are free and equal may not sound very revolutionary to those of us who have lived for generations in liberal democracies, but one can well imagine how it sounded to the kings, popes and noblemen of the seventeenth century!

Liberalism's valuation of the free and equal person has two consequences that are of crucial importance for the liberal attitude to the question of justice for animals. This ethical ideal has had an important effect on the *methodology* many liberals use to determine what justice entails; and it has also had an important consequence on the *content* that liberals give to justice. In terms of methodology, liberals often favour using the device of a social contract to determine what a just society would look like. You will recall from Chapter 2 that the device of a social contract was used by several prominent Western political theorists of the seventeenth and eighteenth centuries. Hobbes, Locke, Rousseau and others all imagined a 'state of nature', where no political community existed, and imagined what political principles individuals in that state of nature would choose when forming a political community. While each of these thinkers had their own views on what the contracting parties would agree to and why, they all agreed that the contract was an important justificatory device. Political arrangements, for these theorists, were justified to the extent that they would have been chosen by the contracting parties. In *A Theory of Justice,* John Rawls revived the use of the contract as part of his justification for liberal institutions. But why do so many liberals consider the device of a contract to be useful in the first place? Well, because liberals value the freedom of individuals, any interference with that freedom, as clearly occurs in the establishment of political rules and rulers, is in a sense to be regretted. But given the benefits of political association, many liberals believe that such authority is justified if it can be shown that it would have been *chosen* by free individuals in some kind of state of nature. As such, the contract reflects the personhood of individuals in that the contracting parties freely choose to be bound by certain rules and principles. Furthermore, the contract also embodies the equality of persons in that it is persons themselves who decide on those rules and principles. Political arrangements are not decided by some mandate from God, nor from the will of a monarch or some other inherently superior individual; instead, they are chosen by those who will be bound by them.

In terms of the content of justice, the liberal valuation of personhood leads liberals to accept pluralism in society. What I mean by this is that liberals take it for granted that individuals have different moral beliefs and different ideas about what a good life entails. Again, to put it in the liberal terminology, liberals accept that people have different 'conceptions of the good'. Liberals accept that for some people, their idea of a good life is watching television and drinking beer, while for others it is a life of pious religious devotion. Liberals believe that the job of politics is not to decide on and enforce the 'correct' conception of the good, but to accept the inevitability of such pluralism, and to allow different conceptions of the good to flourish as far as possible. This is the basis of the liberal principle, 'the right is prior to the good'. Once again, the basis of liberal thought in seventeenth-century Europe is relevant here. After all, liberalism emerged at a time and place not just when traditional hierarchies were being challenged, but also when societies were recovering from centuries of protracted and bloody religious wars. Accepting the fact of religious pluralism then, rather than enforcing the truth by the sword, was something which liberal thinkers of the seventeenth century began to endorse.

The liberal valuation of the individual person, the use of the social contract as a methodology and the acceptance of pluralism in politics have all had important consequences for the consideration of animals in liberal thought. This chapter examines these implications by proceeding in five steps. Firstly, it examines how the use of the contract, particularly in the work of John Rawls, has been used to exclude animals from justice in liberal thought. Secondly, it examines how the liberal promotion of pluralism can lead liberals to permit cruel actions practised against animals. However, in the next two sections, the chapter examines the ways that animal rights proponents have actually used liberal resources to extend justice to animals: firstly, by modifying the social contract so that it applies to animals; and, secondly, by extending the notion of personhood to animals. Finally, the chapter ends by proposing a simple and underexplored way in which liberals can include animals within their political theories. Here it is claimed that when the liberal valuation of personhood *and* welfare is made explicit, liberalism has the resources to extend justice to animals without difficulty. First of all then, let us consider John Rawls's use of the social contract and how it excludes animals.

The exclusion of animals in the Rawlsian contract

You will recall that many liberals favour the use of a contract to justify the rules and principles of political communities because it reflects their valuation of the free and equal person. It is now necessary to examine how that device has been used to exclude animals from the domain of justice. The best way to conduct this examination is through the work of Rawls. The reasons for focusing on Rawls in this way are threefold. First of all, the impact of Rawls on contemporary political theory as a whole has been immense, and any attempt to understand modern liberalism simply must address the work of this great thinker. Secondly, Rawls does explicitly rule out extending justice to animals in *A Theory of Justice*. And finally, a number of animal rights theorists have offered criticisms of Rawls's exclusion of animals that are worth addressing in detail.

To begin with, it is important to note that Rawls sets up his contract in a slightly different way than the contract theorists of the seventeenth and eighteenth centuries. First of all, his equivalent of the state of nature is what he calls 'the original position'. It is important to bear in mind that the original position is not a real place that exists or which has ever existed: it cannot be found on any map, nor at any point in history. Rather, it is a hypothetical device to help us think about which political principles are just. The original position helps us to think about which principles are just by asking us to consider which political principles would be chosen and agreed to by parties within that original position. In order to see which principles would be chosen, Rawls asks us to assume that the contracting parties have certain features. First of all, we are asked to assume that those parties are self-interested and rational. This follows the tradition of previous social contract theorists in that Rawls seeks to formulate just principles not from utopian fantasies about how human beings might be in some ideal future, but from how human beings are. However, Rawls is perfectly aware that parties driven purely by self-interest are not likely to arrive at a set of just principles. For one, if all the parties chose purely on the basis of self-interest then no agreement over political principles could be selected. I would choose principles that advantage me, you would choose principles that advantage you, and no agreement could ever be reached. Secondly, self-interest would obviously not result in principles which we would consider to be very fair. For example, if I selected principles which distributed all of society's resources to white male atheists, most would consider that to be incredibly unfair. After all, being white, male or an atheist are not good reasons to be awarded more resources than anyone else. In Rawls's terms, such facts are 'morally arbitrary', and so not relevant to who should get what in society. For these reasons then, Rawls asks us to imagine that contracting parties are behind a 'veil of ignorance', so that they do not know their own characteristics, talents, conception of the good or social position. As a result, contracting parties do no know whether they are male or female, black or white, rich or poor, clever or stupid, Christian or Muslim and so on. As such, contracting parties choose self-interestedly, but they also choose *impartially*. In effect, the veil of ignorance puts each party in the shoes of every other possible individual in society. So, the way the original position is set up by Rawls, with parties choosing self-interestedly and impartially, ensures that the principles selected will be *fair* and *just*, what he calls 'justice as fairness'.

Rawls believes that the parties would select two principles: the liberty principle, where each individual is granted the right to basic liberties, consistent with the equal right of others; and the difference principle, where social and economic inequalities are only permitted if they bene-fit the worst off, and are attached to positions that are open to all.[1] The reason why Rawls believed that these principles would be chosen should hopefully start to make sense. As a self-interested individual behind the veil of ignorance, it would only be rational to select a principle grant-ing maximum possible basic liberties to individuals. Parties do not know what their conception of the good will be, so it would be wise to build a society which provided for sufficient liberty to allow all conceptions of the good to flourish as far as possible. For example, it would be unwise to choose a policy which persecuted Catholics, when you might end up as a Catholic yourself. Furthermore, as a self-interested and rational individual it would also only be rational to choose a principle which allows for only those inequalities which benefit the worst off. After all, parties themselves might end up as the worst off, and so would want to maximise the position of that group.

Of course, Rawls's defence of these two principles has been extremely controversial. Following the publication of *A Theory of Justice*, Rawls has been subjected to a huge number of criticisms, some of which question the principles themselves, and others of which question his method at arriving them. However, this ongoing debate in political theory is not the one that concerns us. Instead, I want to examine the impli-cations of Rawls's theory for animals. For while Rawls did not present a sustained analysis of the animal issue in *A Theory of Justice* or any-where else, he did explicitly state that animals are excluded from justice, even if his statements to that effect were always expressed in some-what tentative language. For example, of his principles of justice, he writes: 'Our conduct towards animals is not regulated by these princi-ples, or so it is generally believed.'[2] And when it comes to assigning rights to all persons, he writes: 'Presumably this excludes animals; they have some protection certainly but their status is not that of human beings.'[3] So why does Rawls exclude animals from his principles in this way?

Two reasons can be found in Rawls's book. To understand the first, it is necessary to dig a bit deeper into Rawls's understanding of what these principles of justice are for. For Rawls, principles of justice regulate soci-ety, where society is considered as a scheme of social cooperation whose

members can each advance their own good. Any scheme of social cooperation will create burdens and benefits for its members, and according
to Rawls, it is up to the principles of justice to regulate those burdens
and benefits. Important for Rawls then is the idea of *reciprocity*: 'We are
not to gain from the cooperative labors of others without doing our fair
share.'[4] To benefit from the cooperative scheme it is necessary to give
something up to that cooperative scheme. As such, reciprocity demands
that all recipients of justice be *contributors*. Now given that animals do
not contribute to society, in the sense that they do not earn money
or provide other resources of cooperative benefit, it can be argued that
the principles of justice which regulate that society do not apply to
them.

The second reason for excluding animals is Rawls's stipulation that
the characteristics of the contracting parties are the same as those to
whom they apply.[5] For Rawls, contracting parties must be 'moral persons'. In other words, they must be capable of having a conception of
the good, and they must be capable of having a sense of justice. After
all, it is hard to imagine how contracting parties could devise any principles of justice if they themselves did not possess such characteristics.
As such, Rawls argues that '... equal justice is owed to those who have
the capacity to take part in and to act in accordance with the public
understanding of the initial situation'.[6] Because animals do not contribute to the scheme of social cooperation, and because they are not
moral persons then they are not owed justice. Rawls does say that we
have a duty of compassion not to be cruel to animals; but a duty of
compassion is quite different to a duty of justice which can be enforced
by the state.[7]

Of course, both of Rawls's reasons for excluding animals from justice
can be criticised. In the first place, if justice is only owed to those individuals who contribute to the scheme of social cooperation, then justice
must in fact be owed to a great many animals. All those animals on
farms, in research laboratories, in zoos, in our homes and so on make a
contribution to the functioning of society, and should thus be included
as members of a cooperative society. Rawls ignores this fact, and sometimes writes as if animals exist completely outside of human societies.
For example, when writing of the duties of compassion and humanity
we possess to animals, he states:

> I shall not attempt to explain these considered beliefs. They are out-
> side the scope of the theory of justice, and it does not seem possible
> to extend the contract doctrine so as to include them in a natural

way. A correct conception of our relations to animals and to nature would seem to depend upon a theory of the natural order and our place in it.[8]

But of course, our treatment of animals does not depend solely upon a theory of the natural order. Since many animals are part of society, and contribute to society, our treatment of those animals would seem to depend on a theory of the just society.

However, even if Rawls's first basis for excluding animals can easily be dismissed, his second cannot. His second, and most explicit reason for excluding animals, is the fact they are not moral persons. But excluding animals from justice on the basis that they lack moral personality is not without its problems. For one, a good many *human beings* are non-persons: many young infants and severely mentally disabled individuals, for example, possess no conceptions of the good, and have no sense of justice. Does Rawls then believe that his principles do not apply to these individuals? No, he does not. Rawls is keen to maintain that all humans are owed justice, even if they lack moral personality. Rawls maintains that infants are owed justice on the basis that they will one day possess moral personality, and that the severely mentally disabled are owed it on the basis that they possessed it in the past. Rawls views the lack of moral personality of some human beings as something which is 'morally arbitrary'. Such individuals may not have the capacities necessary to be a contracting party, but that is simply because those capacities are temporarily switched off through bad luck. Moreover, Rawls believes that the choice of principles of justice should not be affected by such contingencies: 'Therefore it is reasonable to say that those who could take part in the initial agreement were it not for fortuitous circumstances, are assured equal justice.'[9]

Of course, one large problem emerges from this claim: it is perfectly evident that not all human beings who lack moral personality have it temporarily switched off. Quite clearly, some human beings have mental disabilities that are so severe and permanent that they will never be able to frame and pursue their own conceptions of the good, nor will they ever be able to ruminate and act upon what they consider to be just and unjust. Some humans lack moral personality permanently. Are these individuals to be excluded from Rawls's scheme of justice? Once again, they are not. Rawls recognises that such individuals present a difficulty for his argument, but is very keen to maintain that moral personality – whether possessed presently, in the future or in the past – is a sufficient, but not a necessary condition for being owed justice; at least, that is, if you are a human being. Writing in respect of the moral personality of human beings, he states:

> We cannot go far wrong in supposing that the sufficient condition is always satisfied. Even if the capacity were necessary, it would be unwise in practice to withhold justice on this ground. The risk to just institutions would be too great.[10]

Here then, we get to the nub of the matter. According to Rawls, the principles of justice do not apply to animals because they lack moral personality. However, the principles of justice apply to all human beings, irrespective of their possession of moral personality, because to withhold justice from such individuals would be too great a 'risk to just institutions'.

So what exactly does Rawls mean by this idea of a 'risk to just institutions'? Unfortunately, Rawls does not say any more on the issue. However, Peter Carruthers has helpfully elaborated on the claim in his endorsement of it. Carruthers argues that just institutions would be threatened by excluding humans lacking moral personality in two ways. First of all, he presents a 'slippery slope' argument. Carruthers claims that it is hard to discern precisely who in the human community is a moral person and who is not. As such, if we exclude humans lacking personhood from justice, we are likely to end up excluding some humans *with* personhood. In order to provide robust protection to all human persons Carruthers claims that it would be safer to simply include all humans within the domain of justice. Since, on the other hand, it is easy to discern who is a non-human animal, he claims that no threat to persons emerges by excluding animals from justice.[11] The second way in which excluding human non-persons would threaten just institutions, Carruthers argues, is from its disruption of social stability. Carruthers argues that contractors simply would not be able to comply with any principle which denied justice to human non-persons. He claims that because of their natural bonds of attachment and affection to all humans, including babies and the severely mentally disabled, the only way in which contractors could choose principles which did not result in widespread and prolonged resistance would be through granting justice to each and every human being, irrespective of their capacities. Once again, Carruthers believes that no such threat of non-compliance and resistance emerges by excluding animals from justice.[12]

However, there are a number of important problems with the arguments presented by Carruthers. For example, the slippery slope argument asks us to accept that excluding animals from justice poses no threat to persons. But this can be questioned on two grounds. First of all, there is the possibility that a society which inflicts gross and unnecessary suffering on animals may be more willing to accept gross and unnecessary suffering inflicted on human beings. Clearly this is an empirical claim which requires verification. But Carruthers is wrong to simply ignore the threat to just institutions that the exclusion of animals may cause. Secondly, and perhaps more importantly, Carruthers simply assumes that no animals are persons. He completely ignores the possibility that excluding animals might threaten persons, because some animals *are* persons. Once again, to say that some animals are persons is an empirical claim which needs verification. However, there is great debate amongst zoologists and philosophers over whether the cognitive powers of such animals as the great apes and cetaceans are sufficient to regard them as moral and autonomous beings.[13] To completely ignore such debates is unjustified.

The argument from social stability also raises problems. First of all, there is the point that almost every society throughout history has excluded certain groups of human beings from justice, but has nevertheless remained stable. As pointed out above, the moral equality of all humans is an idea that has only received widespread assent quite recently. It is perfectly possible for societies to be stable, and yet for huge sections of that society to be denied justice, whether persons or not. Secondly, it is unclear why excluding animals from justice would *not* result in widespread resistance and non-compliance. After all, a great many individuals surely believe that animals are owed justice in *some* regard. For all this means is that the state should enforce some restrictions on the way human beings treat animals for the sake of animals themselves. Many would be appalled by and willing to resist principles of justice which said absolutely nothing about the proper limits to our treatment of animals. Once again then, Carruthers's attempt to justify the claim that the exclusion of humans threatens just institutions in a way that the exclusion of animals does not remains extremely dubious.

In all, Rawls's basis for excluding animals from justice is unconvincing. Animals cannot be excluded on the basis that they cannot reciprocate, because many animals obviously do contribute to the scheme of social cooperation. It is also dangerous to exclude animals on the basis that they lack personhood, because then many humans must also be excluded. For as we have seen, arguments which strive to keep all humans included in justice, but which aim to keep all animals out, need to jump through some pretty implausible philosophical hoops to make their case.

The implications of liberal pluralism for animals

As we have seen, Rawls puts great importance on the notion of 'person-hood' as a means by which to make decisions about who merits justice. But we have also seen that Rawls uses it to make decisions about what justice actually entails. After all, the liberty principle embodies the value of personhood by permitting pluralism. That principle grants to every citizen the maximum possible liberty to formulate and pursue their own conceptions of the good. Importantly, Rawls also states that the liberty principle has *priority* over the difference principle, making the contro-versial claim that parties in the original position would not be willing to sacrifice liberty for more resources.[14]

This prizing of pluralism is crucial for liberal conceptions of justice, and the emphasis Rawls puts on it reflects his acceptance of Kantian ideas about personhood. You will recall that Immanuel Kant claimed that human beings possess dignity because they are persons, or 'ends-in-themselves': self-legislating beings who can exercise moral agency and free will. Those entities who cannot exercise moral agency or free will, on the other hand, do not possess dignity for Kant, but are instead mere 'things'. This classification of entities into persons or things had great moral significance for Kant. Kant argued that human beings, as pos-sessors of dignity, should always be treated as ends and never solely as means. Those beings without dignity, however, only have relative value and thus can permissibly be treated as mere means. Many lib-erals, like Rawls, have adopted this Kantian valuation of personhood and used it to claim that states must show their citizens *respect*. That is to say, they have argued that the state must respect the personhood of its citizens, and their capacities to make up their own minds about certain important moral and ethical issues. In other words, these lib-erals claim that the state should leave its citizens to form, revise and pursue their own conceptions of the good with as little interference as possible. As we can see then, for liberals, the capacity for person-hood is not just crucial to the question of who gets justice, but also what that justice entails. For liberals, a just political community should permit a plurality of conceptions of the good to flourish alongside one another; it is a political community with *limited* interference in the lives of individuals.

However, when personhood is used by liberals to both exclude animals from justice and to promote pluralism, it has a detrimental impact on the interests of animals.[15] To explain, consider once again Rawls's writings on animals in *A Theory of Justice*. Some authors have claimed that *A Theory of Justice* can offer strong forms of protection to animals. Ruth Abbey, for example, has argued that Rawls does not grant us license to treat animals just as we please, but imposes upon us an explicit moral duty to show them compassion and humanity. This duty may not be a duty of justice, and it may not grant animals any rights, but Abbey points out that it is a duty nonetheless, and thus can be used to provide strong protection to animals.[16] The problem with this argument, however, is that this moral duty will inevitably come into conflict with Rawlsian justice.[17] After all, the first principle of Rawls's theory of justice is the liberty principle, where each individual is granted the right to basic liberties, consistent with the equal right of others. As we have seen, this liberty principle reflects and encapsulates Rawls's valuation of personhood. The liberty principle allows persons to pursue their autonomous goals, without interference by the state, so long as they do not infringe upon the same basic liberties of others. As such, if my conception of the good requires the daily sacrifice of large numbers of healthy animals to appease the gods, then so long as that does not interfere with the liberties of other persons, such slaughter is perfectly permissible. Rawls may well say that we have a duty of compassion not to be cruel to animals, and so may find my slaughters morally wrong. Perhaps he may even believe that the best, or even true, conception of the good would not encompass such ritual slaughtering. Unfortunately, however, Rawls has no resources to claim that the state should interfere with such slaughters and enforce the duty of compassion. According to Rawls, in order to fully respect personhood, we must let individuals decide for themselves how to live their lives. If their way of life involves the ritual slaughter of animals, or other forms of cruelty to animals, then so be it. By relegating the issue of our treatment of animals to a private moral matter and by promoting the value of pluralism, Rawls condemns animals to all sorts of cruelties that he would find immoral, but not unjust.

A modified contract favouring animals

Despite these tensions between Rawlsian liberalism and the protection of animals, some proponents of animal rights have nevertheless maintained that Rawlsian resources offer excellent support for extending justice to animals. The thinkers Donald VanDeVeer and Mark Rowlands, for example, have claimed that a proper application of Rawls's methodology actually *requires* the inclusion of animals in the domain of justice. To explain, recall Rawls's argument for including within justice those humans such as babies and the severely mentally disabled. Rawls argues that it is not the fault of such individuals that they lack the capacities for personhood. The possession of such capacities is an 'arbitrary contingency' which should not affect the choice of principles by the parties. But if the possession of personhood is an arbitrary contingency which should not affect the choice of principles, then surely to achieve true impartiality we should put it behind the veil, just as Rawls claims we should with other arbitrary traits like race, religion, gender, family background and so on. In other words, on Rawls's own terms contracting parties should be ignorant of whether they will end up as moral persons once in society. If we did put knowledge of personhood behind the veil, justice could be extended to non-persons quite straightforwardly, and without odd arguments about threats to just institutions and so on. After all, contracting parties would not know whether they will end up in society as a young infant, or as an individual with severe mental disabilities, and so would have good self-interested reasons for extending protection to such individuals.

Furthermore, once personhood is behind the veil, it seems logical for species membership to go behind it too. After all, if parties do not know that they will be moral persons, why should they know that they will be human beings? For on reflection, it certainly seems that being of the species *Homo sapiens* is what Rawls calls an 'arbitrary contingency'. It is just a mater of brute luck that any of us is born into membership of a particular species. As such, on Rawls's own terms it would appear that species membership should not affect the choice of principles. Crucially, once species membership goes behind the veil, it would be rational for self-interested parties to choose principles that extend justice to animals, because they might end up as animals in society once the veil is lifted. For theorists such as VanDeVeer and Rowlands true impartiality requires the exclusion of *all* arbitrary contingencies from the contracting process, which necessarily leads to the extension of justice to animals.[18]

However, there are a number of immediate problems with this proposal which require consideration. In the first place, it might be objected that putting species membership behind the veil does too much. Some might claim that it would lead to parties coming up with principles that apply to plants, rocks, electricity pylons or any other type of object! After all, contracting parties could just as easily end up being entities such as these, as they could end up as animals. And yet, theorists have responded to this objection by claiming that it would be rational for parties to choose principles which apply only to entities with *interests*. They claim that parties would not care if they ended up as a plant, rock or electricity pylon because such entities cannot be harmed or benefited. These non-sentient entities are incapable of conscious experience, and thus have no welfare which can be set back or promoted. As such, it is claimed that parties would be motivated to choose principles which apply only to entities with interests, like sentient animals.[19]

A second problem with putting species behind the veil nevertheless remains. And this is the issue of plausibility. Could contracting parties really imagine themselves being animals in society? Becoming an individual of another race, generation or gender is one thing, but could parties really imagine being of a different species? Is this a hypothetical imaginative stretch too far? Perhaps not. After all, it is hard to discern the precise difference between imagining that one will end up as a newborn infant say, compared to ending up as a chimp in a research laboratory. Both are pretty imaginative flights of fancy, to be sure. But once the device of the hypothetical original position and its veil of ignorance are accepted, it is hard to see why putting species membership behind the veil is an imaginative step too far.[20] However, perhaps the problem is not simply imagining oneself as a member of some other species, but choosing principles of justice which apply to other species. How would contracting parties know which distribution of benefits and burdens would be just for a cow, pig, chimp or horse? Obviously this does raise a problem, but it is far from clear that it is an insurmountable one. After all, devising the appropriate principles of justice for fellow human beings is a tricky task, as the continuing debates amongst political theorists are evidence of. And yet, just because it is difficult to devise principles for humans, that does not make it impossible. Perhaps then, the same is true for animals. Furthermore, we do make judgements about how the lives of animals fare quite regularly and without controversy. For good reason we are confident that the life of a factory-farmed chicken is pretty miserable, while the life of a cosseted and loved family dog is pretty good. Given basic facts about their sentience, biological needs and species characteristics, it seems perfectly possible to be able to devise principles that are relevant and appropriate for animals.[21]

However, there is a more fundamental objection to using Rawlsian resources to extend justice to animals which has been made by Robert

Garner. Garner agrees with VanDeVeer and Rowlands in that the logic of Rawls's argument leads to making contractors ignorant of their species. However, Garner disagrees with the basic idea of using Rawlsian resources to extend justice to animals in the first place. Garner's problem with Rawlsian methodology is its reliance on pre-existing intuitions. To understand Garner's objection, it is worth spending a moment to explain that for Rawls, the justification of his theory depends on it being in 'reflective equilibrium'. Rawls believes that principles of justice are justified to the extent that they hang together coherently with what he calls our 'considered judgements'. To explain, recall the way in which the original position is set up. Parties are self-interested, but that self-interest is tempered by the veil of ignorance which ensures impartiality. According to Rawls, this veil models our considered judgements about fairness. For Rawls, many of us possess the basic considered belief that such things as race, religion, gender and so on are irrelevant to who gets what in society. Now Rawls argues that principles of justice can be ranked in accordance with their conformity with this considered judgement about fairness. This is not to say that these judgements are fixed and must always be complied with, but it does mean that when principles conflict with these judgements we face a choice: if the judgements are so basic, we may have to modify our principles; but if the principles have such explanatory power, we may have to modify our considered judgements. Rawls claims that when devising our theory, we should revise upwards and downwards in this way until our considered judgements and principles are in 'reflective equilibrium', with the theory hanging together as a coherent whole. Rawls believes that his principles of justice are in reflective equilibrium with our considered ethical judgements about fairness.[22]

According to Garner, however, this justificatory process of reflective equilibrium renders the contractual argument all but redundant.[23] Garner claims that the bulk of the justificatory work in Rawls's theory comes not from imagining what contractors in the original position would choose, but from certain pre-existing normative assumptions about fairness. Moreover, if those assumptions are altered, radically different principles follow. This, Garner believes, explains the different views about the extension of justice to animals between Rawls and animal rights theorists like VanDeVeer and Rowlands. For Rawls, excluding animals from justice conforms to his pre-existing convictions about fairness: for him, fairness relates only to human beings, because they alone are persons. While for animal rights theorists, including animals in justice conforms to *their* pre-existing convictions about fairness: for them, fairness relates to all animals in so far as they are sentient.[24] Because of the importance of considered judgements to the justificatory process, Garner is sceptical about whether Rawlsian resources offer good grounds for extending justice to animals. In the end, everything depends on one's pre-existing convictions about fairness. Perhaps then, it would be better to simply make these convictions explicit and offer good arguments in support of them. For example, perhaps VanDeVeer and Rowlands would be better off explaining and defending the view that it is the capacity for sentience that makes one a legitimate recipient of justice, rather than fiddling around with the details of the original position and veil of ignorance.

A modified conception of personhood favouring animals

Rather than attempting to modify the social contract, other proponents of justice for animals have instead attempted to offer direct arguments for why animals count. Moreover, some of those thinkers have drawn on the liberal notion of personhood to make their case. Indeed, several animal rights theorists have argued that animals *do* possess the necessary characteristics for personhood. Firstly, and as noted above, some theorists have pointed to the capacities of such animals as the great apes and cetaceans, and have argued that it is at least *possible* that such creatures are self-legislating and moral creatures.[25] Hence they have questioned the liberal assumption that only human beings can be persons, and thus that only human beings can merit justice. However, let us leave this particular ascription of personhood to animals to one side. For not only does it rely on controversial questions about the natures of animals' cognitive abilities, but even if it were correct, it would not provide for a very expansive extension of personhood to animals. For in fact, several theorists have argued that personhood can be extended to the vast *majority* of sentient animals quite straightforwardly. They have argued that what is crucial about personhood is not the capacity for moral or rational reflection, but the capacity for *agency*, a capacity which a great many sentient animals certainly possess. For example, Tom Regan has argued that many sentient animals are in fact autonomous agents. Regan admits that it is highly unlikely that many animals are autonomous in the Kantian sense: as having the ability to think through the merits of acting in one way or another, and to make a decision based on those deliberations. However, Regan goes on to write:

> But the Kantian sense of autonomy is not the only one. An alternative view is that individuals are autonomous if they have preferences and if they have the ability to initiate action with a view to satisfying them.[26]

As we saw in the previous chapter, Regan argues that this capacity to initiate action and pursue preferences is one of the necessary conditions for being a 'subject-of-a-life'. And all entities who are subjects-of-a-life possess 'inherent value', a value which cannot be reduced simply to their usefulness to others. As you will recall, Regan claims that these types of agent have a right to respectful treatment, which requires that they never be treated merely as means.[27] As should be clear then, Regan adapts the Kantian source of value to reach a Kantian conclusion for animals. Regan argues that it is 'preference autonomy', rather than 'Kantian autonomy', which is morally relevant, and it is the capacity for preference autonomy which entitles the possessor to a right to respectful treatment. Steven Wise and Evelyn Pluhar are two other theorists who have followed the logic of this argument, by also extending personhood and basic rights to animals on the basis of their agency.[28]

But what does a right to respectful treatment actually entail politically? If justice were to be extended to animals on the basis of their agency, how would political communities meet their obligation to treat animals with respect? Gary Francione has drawn on the work of Tom Regan to argue that the right to respectful treatment essentially requires that animals stop being used for human ends, irrespective of whether that use causes suffering.[29] Francione writes: 'The use of animals for food, sport, entertainment, or research involves treating animals merely as means to ends, and this constitutes a violation of the respect principle.'[30] Furthermore, Francione writes that such a change cannot take place while animals continue to be considered the *property* of human beings. Francione contends that property cannot possess rights, and since animals are currently considered to be property in the vast majority of political communities, animals can possess no rights. As such, according to Francione, Regan's moral injunction to treat animals with respect requires political communities to recognise the personhood of animals by abolishing their property status.[31]

However, there are a number of problems with these attempts to extend personhood to animals. For even if we concede that animals are agents and pursue desires, such volition is not what grounds the liberal valuation of *personhood*. In the first place, the claim that because animals are agents, they should always be treated as ends-in-themselves and never solely as means is very problematic. As Regan himself admits, agency is not the same as Kantian autonomy. The vast majority of animals are not ends-in-themselves: they cannot self-legislate and act morally. Since animals are not ends-in-themselves, we are left wondering why these animals should always be *treated* as ends-in-themselves. If most animals are not autonomous beings who can formulate, revise and pursue their own conceptions of the good, why do we have an obligation not to use them as means to our ends? Of course, the answer might be that using animals as means in this way – as we do on the farm or in the laboratory, for example – causes them to suffer. That is undoubtedly true. But in that case, our obligation not to use animals is merely derivative, derived from our obligation not to cause them to suffer. Regan and Francione contend that it is the use of animals per se that causes the harm, irrespective of whether that use causes suffering. But since animals are not ends-in-themselves, autonomous beings able to frame and pursue their own goals and aims in life, why does the painless use of them cause them harm? The response might be that by using animals in these ways, we necessarily restrict their *agency*, thus causing harm. But is that really the case? Many of us who keep pets, or who ride horses for recreation, will find it hard to believe that each and every use of animals involves a harmful restriction of their agency. After all, keeping a non-autonomous animal as a pet is surely quite different from keeping an autonomous human being as a slave.[32]

Furthermore, the claim that the property status of animals necessarily impedes achieving justice for them can be questioned on at least three grounds. First of all, we can question whether abolishing the property status of animals will necessarily lead to their better treatment. After all, it has been pointed out that many wild animals are regularly treated cruelly, despite not being the property of private citizens.[33] Secondly, we can question the basic assumption that once something is owned, it cannot then possess rights. After all, corporations are very much owned, but still possess certain rights and responsibilities. Indeed, many would argue that animal welfare legislation does, as a matter of fact, confer rights on owned individual animals in many states.[34] Finally, and perhaps most importantly, just because animals are property, that does not mean that human beings can do what they like to them. It is perfectly possible for states to continue to recognise animals as property but also to impose strict restrictions on what owners may permissibly do to them. After all, property owners do not have an inviolable and absolute right to do exactly what they like with their property under any legal regime. Once this is recognised, it may be perfectly possible to own animals and to nevertheless treat them justly.[35]

Valuing personhood *and* welfare

Given these problems with the modified contract and the modified notion of personhood, must we conclude that liberalism offers scant resources for extending justice to animals? Not necessarily. For I want to point out that most liberals do not believe that personhood is *all* that is of value. Nor do they believe that the state's *only* function is to respect the personhood of its citizens. Instead, most liberals value personhood *and* welfare. For example, consider Ronald Dworkin's interpretation of Rawls's theory of justice as fairness:

> ... justice as fairness rests on the assumption of a natural right of all men and women to equality of concern and respect, a right they possess not by virtue of birth or characteristic or merit or excellence but simply as human beings with the capacity to make plans and give justice.[36]

Here Dworkin argues that justice as fairness rests on the assumption that all human beings have a right to be treated with equal concern and respect by the state. Whether Dworkin is right in his interpretation of Rawls is not our concern. What matters for us is his endorsement of the principle of 'equal concern and respect'. This principle has important implications for animals, because it embodies a valuation of personhood *and* welfare.

To explain, it is worth saying something about what the principle actually involves. In the first place, the ideal of equal *respect* derives from the familiar Kantian valuation of personhood. Because persons are autonomous, self-determining beings, the state should respect the choices that individuals make for their own lives as far as that is possible. However, the ideal of equal *concern* relates to something quite different. It derives from the notion that our interests are not exhausted by our autonomy or personhood. Because individuals are sentient, needy beings, the state should show concern for the basic needs and welfares of its citizens. Now while Dworkin restricts his principle of equal concern and respect to 'human beings with the capacity to make plans and give justice', it is obvious how it can be extended beyond such narrow confines. If states should have as their focus respect *and* concern, personhood *and* welfare, then surely justice is owed not only to persons. Beings without personhood, but with a welfare, such as young infants, the severely mentally disabled and animals, should also be of *concern* to them. In other words, the principle of equal concern and respect demands that liberal states be interested and concerned about the suffering and needs of sentient non-persons, including non-human animals.

Furthermore, once it is recognised that liberal states have an obligation to both respect personhood and to be concerned about well-being, it becomes evident that the latter can sometimes take priority over the former. In other words, we must recognise that liberty can sometimes be legitimately restricted for the sake of welfare. For example, we are familiar with the liberal claim that individuals should be free to pursue their own conceptions of the good, *as far as possible*. Of course, the 'as far as possible' caveat is important. Individuals cannot pursue conceptions of the good which involve setting back the fundamental interests of others. In other words, liberals believe that the state should leave us free to do as we like so long as we do not harm others. This familiar liberal refrain derives from John Stuart Mill's famous 'harm principle': 'That the only purpose for which power can be rightfully exercised over any member of a civilized community, against his will is to prevent harm to others.'[37] And to reiterate, harm does not just mean imposing on the freedom of others; it also concerns imposing on their well-being. After all, we all agree that the state should restrict the liberty of others when it comes to assault and the infliction of other physical injuries. Crucially, this harm principle can also be applied to animals. For animals are also subjects of well-being who possess fundamental interests. As such, once personhood *and* welfare are recognised to be of value, liberals can coherently place limits on the freedom of individuals out of concern for animal well-being. To return to the example given above then, contrary to Rawls, liberals can argue that my daily practice of slaughtering animals is wrong and should be outlawed, irrespective of the fact that this will impose on my liberty. After all, it can be claimed quite plausibly that the harm I inflict on the animals is sufficient to warrant the state interfering with and putting a stop to my practices.

The great difficulty with this approach, of course, is working out just when and why liberty can be restricted for the sake of welfare. For example, what if my slaughter involved just one animal, just once a year, and was absolutely integral to my religious beliefs? Should the state then legitimately step in to ban it? And if liberty can be subordinated to welfare in the case of religious slaughter, should it also be subordinated when it comes to practices such as hunting, horseracing, meat-eating and countless other practices? To many liberals, the idea of the state interfering with such individual practices is anathema to the value of the free and equal person.

Irrespective of these concerns about the appropriate balance between liberty and welfare, it is evident that most liberals value both. Liberals can value personhood, without claiming that personhood exhausts justice, and without claiming that only persons are owed justice. Once liberals recognise that personhood *and* welfare count, then animals can coherently be recognised as legitimate beneficiaries of justice.

Conclusion

In this chapter we have examined the relationship between liberal thought and the obligations of political communities to animals, primarily by focusing on the work of John Rawls. Rawls revolutionised liberalism and political theory in the 1970s by attempting to justify liberal political arrangements without relying on utilitarian calculations, and instead by taking the separateness of persons seriously. This liberal valuation of the person led Rawls, like so many other liberals, to favour using the device of a social contract to help determine what justice is, and to give pluralism a prominent place in that account of justice. However, as we have seen, this use of the contract and this prizing of pluralism have had important consequences for animals. Both can easily be used as a basis to exclude animals from justice. In fact, Rawls himself explicitly proposed that animals are not owed justice. Surprisingly, however, several animal rights theorists have attempted to modify these liberal resources in order to extend justice to animals. Some have claimed that animals can benefit from a social contract if parties to that contract are made ignorant of their species. Others have claimed that animals do possess the important characteristics of personhood, such as the capacity to possess and pursue desires. However, we have seen that such modifications are problematic. Rawls's use of the contract is really only as a device to model what is fair, which in turn is based on a prior conviction about the value of *persons*. As such, it cannot be readily applied to sentient animals. For as we have seen, for liberals, personhood is something more than simple volition: it usually means the ability to formulate and pursue one's own conception of the good.

However, many liberals would claim that the freedom of the individual person is not the only thing that they value; welfare counts too. Once concern for individuals' well-being enters the picture, the possibility of opening the door to justice to animals becomes available to liberal thinkers. Quite obviously, if liberals are to value personhood and welfare, they will require some convincing means of settling disputes when the two values conflict, as they regularly do. We have already touched upon some examples of such conflicts in this chapter, and in the next we will consider even more. For in the next chapter we will examine whether states should permit religious and cultural minorities to employ *different* standards of animal welfare to the rest of society. In this context, it is not only the values of liberty and welfare that need to be balanced, but also that of *community*.

5
Communitarianism and Animals

We saw in the last chapter that John Rawls published his *A Theory of Justice* as a response to the dominant utilitarian ideas of the time. Rawls wanted to offer a defence of liberal democratic institutions that was not contingent on overall welfare and which took the so-called separateness of persons seriously. While Rawls's two principles were not suddenly heralded and adopted by the political leaders of the time, in terms of its impact on political theory, Rawls's project can be viewed as something of a success. After all, numerous alternative liberal theories emerged after the publication of Rawls's book, and it is fair to say that liberalism has been the dominant political theory in the Anglo-American world ever since. Perhaps unsurprisingly, however, liberalism has not been without its critics. One of the most important challenges to Rawls's thought, and to liberalism more generally, emerged in the 1980s from a group of thinkers who have been labelled 'communitarians'. While all of the theories examined in this book come in different shapes and sizes, with all having their own internal disputes, it is perhaps fair to say that such diversity is most pronounced in the case of communitarianism. For one, some of those thinkers most famously associated with communitarianism – Michael Walzer, Michael Sandel, Alasdair MacIntyre and Charles Taylor – have resisted the tag.[1] And it is certainly evident that the writings and recommendations of these important thinkers vary considerably and in important ways. However, communitarianism is rightly considered to be a meaningful and recognisable position in political theory on the basis of its distinctive core.

The core of communitarianism, and a core which is endorsed by all the theorists who have been grouped under the label, is its endorsement of a politics of the common good. In other words, communitarians

believe that political communities should frame their political principles, institutions and policies around their shared communal goals and
values. According to communitarians, liberals take the 'separateness of
persons' too far. They blame liberal individualism for many of the ills of
modern society: the breakdown of family life, mindless consumerism,
increasing rates of mental illness, rising crime rates, increasing political
apathy and a general decline in communal activity. Communitarians
thus urge for a politics which places less emphasis on the rights and freedoms of individuals, and more on the goods and values of communities.
In direct opposition to liberals then, communitarians give priority to
the communal good over individual rights; they resist the liberal refrain
that the right is prior to the good. Communitarians also take issue with
the liberal claim that the state should step back from important moral
questions, allowing individuals to pursue their own conceptions of the
good with as little interference as possible. For communitarians, the
state should be *active*. They claim that it *is* the job of the state to take
moral positions, and to promote and foster communal values and goals.
On this basis, communitarians would argue that it is right and proper,
for example, for the state to discourage lives stuck on the sofa, swilling
beer and ordering useless goods from various television shopping chan-
nels, and to actively encourage lives of political activity that promote
communal values. In many ways then, communitarians look back to
the ethics of Aristotle which, as we saw in Chapter 2, emphasise man's
shared communal and political capacities. Communitarianism claims
that a virtuous life for human beings exists, and that it is the state's job
to promote it.

But while communitarianism's differences with liberalism are obvious, some might claim that its differences with utilitarianism are less
clear. After all, isn't utilitarianism also interested in a politics of the
common good: the promotion of overall welfare? Yes it is, but it is
important to bear in mind how utilitarians define the common good.
For utilitarians, there is one single measurement of the common good:
overall welfare. According to communitarians, however, the common
good cannot be boiled down to scientific calculations of utility. Instead,
the common good will depend on the type of society that is under
discussion, and the shared traditions and customs that it has. To put
it in the terms of political theorists, communitarianism is *particularistic*, while utilitarianism is *universalistic*. As such, for communitarians
there is no single common good applicable to all societies, but rather
a range of common goods particular to the political communities in
question.

So what does all this mean for the treatment of animals? What do communitarians have to say about how political communities should treat animals? Well, it is important to re-emphasise the diversity of communitarian thought that I discussed at the outset of this chapter. There is no single communitarian position on the treatment of animals, and in fact most communitarian writers have little to say on the obligations political communities have towards them. Nevertheless, the core tenet of communitarian thought – the promotion of the politics of the common good – has important implications for the treatment of animals. Those implications are explored in this chapter in five stages. First of all, the chapter proposes a possible communitarian theory of justice for animals. Communitarianism has the potential to put forward a theory of justice for animals that awards them extremely strong protection. For instance, in societies which care greatly about the interests of animals – the UK could be given as an example – communitarians would surely argue that the state should pursue strong policies of animal protection, and should pursue those policies even when they restrict the rights and liberties of individuals. However, a communitarian theory of this kind does raise difficulties, and the remainder of the chapter examines some of the most important. Firstly, the problem of particularism is addressed. For if justice should be extended to animals because some societies value animals highly, what about the protection of animals in those societies which do not? Secondly, the problem of determining the 'shared values' of a society is examined. After all, don't communities contain quite different values and goals, including quite different attitudes to animals? Thirdly, the problem of partiality is addressed. For if animal protection depends on the shared values of societies, won't unpopular animals be left vulnerable? And finally, the chapter asks which community's values and goals states should seek to promote. After all, it is evident that states contain a number of different communities, each of which often values animals in quite different ways. This chapter explores each of these issues in turn.

A communitarian theory of justice for animals

As we have just discussed, communitarians believe that the state should actively strive to promote communal values. This is unlike liberals, who want the state to remain neutral on moral matters and only intervene in individuals' lives to uphold their rights and freedoms. Now, as we saw in the last chapter, the liberal valuation of persons and their individual freedoms can be bad news for animals. Because the default liberal position is non-interference in the autonomous choices of persons, many liberals are reluctant to outlaw certain practices which cause grave suffering to animals, even if they themselves believe them to be cruel or immoral. Recall that Rawls himself believed that we have good moral reasons not be cruel to animals, but also believed that animals should be excluded from considerations of justice. This non-interference by the state leaves animals vulnerable. Under Rawlsian liberalism, we are personally free to condemn practices which cause animals to suffer, but there are scant resources for the state to coercively intervene to outlaw those practices. Perhaps then, communitarianism, which urges states to be interventionist and to promote communal goals and values, offers better resources for extending strong forms of protection to animals. In this section, I want to suggest one such way that communitarianism can extend justice to animals, using the example of British society.

First of all, it is evident that the British have chosen to extend justice to animals by having an active state that outlaws various forms of cruelty to animals. After all, the first modern piece of animal welfare legislation in the world, the so-called 'Martin's Act', was British, and passed way back in 1822. And it is fair to say that because of the lobbying and campaigns of the British people, the UK has some of the strongest farm animal welfare legislation in the world, often imposing much stricter standards than it is obliged to under EU law. For example, veal crates were made illegal in 1990, some 17 years prior to an EU ban; sow stalls have been banned in the UK since 1999, an EU ban not due until 2013; the production of foie gras is entirely banned in the UK, unlike in many of its neighbouring countries; and the Animal Welfare Act of 2007 not only imposes an obligation on animal owners in England and Wales not to cause unnecessary suffering, but it also imposes on them a duty of care to take positive steps to ensure that the interests of their animals are met.

However, in spite of this relatively progressive legislation, it is nevertheless possible to claim that UK law fails to live up to the opposition to cruelty felt in Britain. After all, the intensive farming of animals continues unabated in the UK, no doubt because of the reluctance of governments to interfere with the rights and liberties of farmers, agribusiness and consumers. But this intensive farming causes huge amounts of suffering, suffering which the British people oppose and reject when they are made aware of it. For example, a campaign concerning the plight of battery chickens in 2008 by two British chefs led to a 50 per cent rise in the sale of free-range chickens.[2] This opposition to such suffering is in line with the British people's traditional affection for animals, an affection which is regarded by other nations as eccentric to say the least. For instance, reflecting on the characteristics that single out the English from other nations in his 1947 book, *The English People*, George Orwell wrote:

> Although its worst follies are committed by upper-class women, the animal cult runs right through the nation...several years of stringent rationing have failed to reduce the dog and cat population, and even in the poor quarter of big towns the bird fanciers' shops display canary seed at prices ranging up to 25 shillings.[3]

Now in light of such attitudes, it is possible for communitarians to argue that the British state should take a more active stance in promoting this communal opposition to cruelty. For example, it is possible to argue that the UK government should take a clear moral stance on the cruelty of factory farming; a stance which is more consistent with the attitudes and values of the British people. This position would entail an outright ban on all intensive methods of farming, and thus an end to some of the cruellest ways we treat animals. Such a ban would indeed interfere with the rights and liberties of individuals, but it would nevertheless be in keeping with the communal good of British society, and thus would be supported by a communitarian political theory.

The remainder of this chapter identifies and explores four important difficulties faced by this type of communitarian theory of justice for animals.

Particularism

The first problem with the communitarian case for animal justice outlined above is that it leaves the extension of justice to animals entirely *contingent* on the values and norms of the society in question. So while this theory might extend justice to animals in a country like the UK, it would offer no such protection to animals in societies that have no concern for the interests of animals. Indeed, in some societies it could even leave animals more vulnerable, by rolling back anti-cruelty legislation in the name of societal indifference towards them.

In response, many communitarians would simply take this objection on the chin: they would accept that their theory is particularistic, but see that as a virtue, rather than a problem. For example, Michael Walzer has written that '...a given society is just if its substantive life is lived in a certain way – that is, in a way faithful to the shared understanding of its members'.[4] For Walzer, in order to show respect for individuals we need to respect the different meanings they attach to goods. For example, in some societies like our own, the consumption of cats and dogs is taboo, whereas in others such consumption is perfectly permissible. To impose our particular standards and values in relation to eating cats and dogs, on this line of reasoning, is to fail to show respect for the standards and values of others. They have constructed their relations with cats and dogs in one way, and we have constructed them in another. On this view, it is unjust to say that they are wrong and that we are right: each society should be free to make those decisions for themselves. On similar lines then, communitarians can argue that it is entirely proper for societies which are strongly opposed to animal suffering to extend justice to animals, say by outlawing intensive farming practices; but it is also entirely proper for those societies which do not have much concern for animals to be free to use them. In other words, there is no single theory of justice for animals that is the same in all places at all times.

The grave problem with this kind of ethical particularism, however, is that it closes off critical reflection on all sorts of cruel practices that surely warrant debate. After all, it seems perverse to be *entirely* uncritical of *all* ethical norms of *all* societies. For these norms are not just trivial matters of manners, etiquette and custom, but have real consequences for the well-being of individuals. For example, we are all aware that many societies have practices and values that are hugely discriminatory towards women, and to religious, national and ethnic minorities.[5] Many of these practices prevent such individuals from accessing education, taking up certain jobs and possessing equal status under the law. Must these social practices and values, no matter how appalling, be totally immune from criticism? This does seem odd in the extreme. In fact, most communitarians are prepared to acknowledge the existence of *some* universal values. However, they do want to maintain that those values can be derived from within the particular communities themselves. Bhikhu Parekh, for example, has argued that while some universal values do exist, they are defined and realised differently by different societies.[6] He has argued that values such as 'sexual equality' are not blunt instruments with which to measure different peoples and their practices, but are instead values which need to be understood in a culturally sensitive manner. Perhaps then, the concept of 'animal justice' is a concept and value that can be found in all societies, but a value which is practised and realised quite differently across them.

But there is an important difficulty with this kind of 'contextual universalism'. For we need to consider just how flexible a culturally sensitive interpretation of a universal standard can be, while still remaining loyal to that standard. Clearly, not just anything can be labelled as a legitimate culturally sensitive interpretation of a universal standard. For example, suppose we put forward a universal principle of non-discrimination over sexual orientation. Let us then imagine a political community which not only discriminates against gay people, but actually outlaws homosexuality. Quite simply, it would be perverse to regard this state as having a different *interpretation* of the principle of non-discrimination over sexual orientation. Rather, that community believes in and promotes a *totally different principle*. And this applies in the case of animals too. Some societies clearly have values and practices which are totally indifferent to the suffering of animals. We just need to take a look at certain societies' practices relating to hunting, whaling, slaughter, animal sacrifice, farming, animal fighting and so on to appreciate this obvious fact. If communitarians wish to promote the shared values of societies, then presumably they should support policies in those societies which sanction all sorts of cruelties to animals.

And yet, it is not necessary for communitarians to be entirely uncritical when reflecting on the practices of societies. After all, and as the next section shows, all societies have a diversity of norms and practices. It is thus possible for communitarians to draw on the norms and practices of society that they find the most sympathetic, and to argue that these are the values which the community should foster and promote. For example, it is often argued in the West that the Chinese have a particularly cruel attitude to animals. Practices such as bear farming, where bears are confined in tiny cages so that bile can be extracted from open wounds in their stomachs, are horrific and shocking. But they are also shocking to many Chinese. This shouldn't be surprising, because China does have a tradition of showing compassion to animals. For example, Chinese Buddhism has long been associated with vegetarianism. The monk Dao Xuan (596 AD–667 AD) is famous for removing the lice from his body and placing them where no one could tread on them. While another Chinese monk of more recent times, Xu Yun (1840–1959), actually preached to the animals, much akin to St Francis of Assisi.[7] Communitarians can point to these particular norms and traditions in Chinese society and argue that these are the values which should be fostered. As such, communitarians can advocate for a politics of the common good, and also take a stand on the type of common good that should be promoted.

But this proposal still faces an important problem. The communitarian who uses this argument needs to explain *why* it is that one set of values should be promoted over another. For example, why should China promote the value of compassion to animals, rather than indifference, when both norms exist in Chinese society? Because of the particularism of the theory, a communitarian cannot appeal to some universal norm like overall utility or the inherent value of individual animals to justify their preference. Instead, the communitarian must return to the particular characteristics and norms of the society in question. As such, the communitarian would have to show that the value of compassion to animals is somehow more *authentically* Chinese. Clearly this raises difficulties. For in many cases, proving such authenticity might be impossible, and in cases where such authenticity can be found, those norms might advocate undesirable forms of cruelty.

Defining social values

The above-mentioned difficulty in finding the 'authentic' set of norms of a society is the second problem with the communitarian case for animal justice. For the problem with framing policies around the shared values of a society is determining what those values are. This is much more difficult than my superficial and sweeping discussion of British attitudes to animals suggested. For what is the real British relationship with animals? This is extremely difficult to discern. First of all, there is the problem that societal attitudes evolve and change over time. For example, prior to the nineteenth century, the British were actually regarded by their continental neighbours not as sentimental cat-worshippers, but as amongst the countries *least* affectionate towards their animals. This was because of the prevalence and popularity of such practices as bear-baiting, badger-baiting, dog-fighting and cock-fighting.[8] But if societal attitudes are constantly shifting like this, just which ones are we to take as authentic? After all, communitarians often place great importance on the *traditional* customs of communities. So perhaps any such change in those customs is to be regretted and prevented, as far as that is possible. The problem with this proposal, however, is that societal change is inevitable. Any attempt to prevent societal norms from changing would require an incredibly authoritarian, repressive and insular political regime, which would not only be morally outrageous, but also doomed to fail in the long term. Perhaps then it would be better to take as definitive the values that a society currently has. But as pointed out above, what if those values lead to unquestionable cruelty and discrimination? It remains unclear why individuals should be made to suffer terribly just because they live in societies that have norms and practices which sanction such cruelty.

Of course, some might want to claim that British attitudes are gradually moving towards greater affection for animals, and away from support for practices that are cruel. As such, they might argue that it is these evolving norms that should be taken as definitive. However, such a general trend towards more humane attitudes is difficult to establish definitively. For example, polls conducted by IPSOS-MORI between 1999 and 2009 suggest that public acceptance of animal experimentation has in fact *risen* significantly.[9] While a 2007 survey by DEFRA found just 5 per cent of Britons described themselves as vegetarian or vegan.[10] While this is an increase from polls conducted in the 1980s, it shows no increase from those conducted in the mid 1990s.[11] It can reasonably be argued that if Britons were truly opposed to the suffering of animals, then many more than this would be against the practices of research laboratories and intensive farming.

The second problem with determining the authentic British relationship with animals is the obvious fact that different people in the UK have different attitudes to animals. It is just plain wrong to say that there is some kind of love affair with animals amongst *all* British people. George Orwell was English, and yet scathing of the English affection for animals. And he is by no means an exception. Look, for example, at the furore caused by the Hunting Act of 2004, which banned hunting with hounds, as used in traditional forms of fox-hunting. Clearly there were serious disagreements amongst the population over our proper obligations to animals in this regard. Some claimed that hunting with hounds was a traditional rural pursuit that hunters should be left to pursue without interference. Others claimed that it was a barbaric hangover from the days of badger-baiting and cock-fighting. Proponents of hunting claimed to speak on behalf of the traditional values of those dwelling in the British countryside. Opponents of hunting claimed to speak in accordance with the British tradition of enacting humane and progressive animal welfare legislation. So which side can be claimed to represent the authentic voice of the British? It appears that both sides represent equally authentic voices. In truth then, it is surely the case that Britain has a plurality of attitudes towards animals; the British have no single shared value in relation to animals.

Partiality

The third problem with promoting animal justice on the basis of shared societal values is not only that people have different attitudes to animals, but also that they have different attitudes to *different animals.* In other words, communities often have affection for some animals, but not for others. This kind of partiality leaves some animals extremely vulnerable under a communitarian theory of animal justice. For example, a communitarian theory might leave domesticated pets extremely well protected, but leave wild animals without any significant protections. A pet rat might thus be due all sorts of rights under state legislation, but a wild rat might be left at the mercy of the pest and vermin exterminators. This is a long way off considering the interests of creatures equally, as mandated by such impartial theories of animal justice proposed by Peter Singer.

However, and as was the case with debates over particularism, many communitarian thinkers would regard this ethical partiality as a virtue of the theory, rather than a flaw. For example, Mary Midgley has argued that attempting to formulate our obligations using the kind of strong impartiality that Singer proposes is absurd. According to Midgley, nearness and kinship matter to morality:

> The special interest which parents feel in their own children is not a prejudice, nor is the tendency which most of us would show to rescue, in a fire or other emergency, those closest to us sooner than strangers. These habits of thought and action are not unfair, though they can probably be called discriminatory. There is good reason for such a preference. We are bond-forming creatures, not abstract intellects.[12]

But once Midgley opens the door to partiality in her ethical theory, it becomes necessary to ask whether human beings can legitimately put their own interests above those of other species. And indeed, Midgley argues that they can: her advocacy of partiality extends not merely to those animals that we prefer, but also to the species that we prefer. As such, Midgley argues against speaking in a very whole scale and radical way about the equality of all animals. She claims that to equate speciesism with racism is quite wrong: 'This way of thinking is hard to apply convincingly to locusts, hookworms and spirochaetes, and was invented without much attention to them.'[13]

But Midgley's point is not to use partiality to completely exclude animals from ethical consideration. She believes that we should grant strong protection to animals, but she also believes that we can do so within a framework that allows for partiality. Midgley provides three reasons for such a claim. Firstly, she argues that the interests of humans and animals are rarely so at odds that it is necessary to prioritise one set over the other.[14] For example, becoming a vegetarian does not require us to negate our obligations to our friends, family and fellow nationals. Secondly, she claims that nearness is not all that matters. Other considerations can sometimes override our obligations that are based on ties of kinship.[15] For example, a child might have an obligation to provide a birthday gift for his father, but that need not extend to purchasing him his request of his favourite foie gras! Finally, Midgley argues that we also have ties of kinship with other animals. In fact, she argues that we are members of a 'mixed community' with non-human animals, and thus have obligations to them based on nearness.[16] As such, Midgley wants to provide strong protection for animals within a theory that puts the interests of those closest to us – including humans – first.

But there are two immediate problems with Midgley's theory. In the first place, if it is okay for humans to put humans first on the grounds of kinship, why cannot whites put the interests of whites first, or men the interests of men first, on the same basis? Midgley considers this objection, but dismisses it, arguing that while race and gender are irrelevant to what we owe individuals, species is absolutely crucial. After all, in order to know our obligations to an animal it is crucial that we know whether it is a human, a dog, a flea and so on. However, Midgley argues that it is not crucial that we know their race or their gender.[17] But Midgley seems to simply be wrong on this point. After all, and as Elisa Aaltola has pointed out, sometimes we *do* need to know an individual's race or gender in order to ascertain our obligations.[18] Certain obligations relating to health, reproduction, historical injustice, non-discrimination and so on can all vary depending on the type of human being we are talking about. Furthermore, and perhaps more importantly, just because species is relevant to the *content* of our obligations, that does not mean that it is relevant to the *prioritisation* of our obligations, as Midgley wants to claim.[19]

Secondly, we should be very cautious about translating the uncontroversial partiality we show in our personal lives to partiality in the policy prescriptions made by states. For it is important to bear in mind a distinction that political theorists make between so-called first-order and second-order partiality. First-order partiality relates to the preferences we have for our friends, family and pets in our everyday actions. We give more attention, time, resources and ethical concern to those closest to us, and few would argue against that. However, it is not evident that this partiality should extend to the second-order level, where policy decisions are being made at a state level. For obviously it is only fair that these second-order decisions are as impartial as they can possibly be, so that resources, opportunities and privileges are awarded by state officials without reference to their partial sentiments. As such, it is quite possible, and indeed proper, that we can be in favour of first-order partiality and second-order impartiality.[20]

However, perhaps this is too quick. After all, it is evident that resources are scarce, and often a state will have to choose policies which inevitably favour one set of individuals at the expense of another. Given this inevitability, perhaps partiality provides a reasonable way of making such prioritisations. Perhaps then, there is still a role for partiality to play within a communitarian theory of justice for animals. This is certainly the claim of J. Baird Callicott. Callicott argues that each of us belongs to a number of different communities: family, neighbourhood, state, humanity, mixed community and biotic community. Callicott's claim is that each of these communities has a different structure and thus places different moral requirements upon us.[21] For example, Callicott claims that our obligations to wild animals in the 'biotic community' are quite different to our obligations to domestic animals in the 'mixed community'. He claims that our duties to wild animals are based solely on the flourishing of the biotic community; that is to say, they are based on the preservation and integrity of the natural environment in which they reside. Hence culling overpopulated or non-native wild animals is quite permissible and even obligatory under Callicott's theory. However, Callicott claims that our obligations to domesticated animals are quite different. He argues that we entered into a kind of contract when animals were domesticated and made part of the so-called mixed community. This contract entailed that we could use and even slaughter domesticated animals, provided that we grant them appropriate pastoral care. Callicott argues that factory farms are a gross violation of this contract and are thus wrong as such.

Crucially, under Callicott's theory, not only are the obligations we have in respect of these various communities different in content, but they are also different in priority. Callicott claims that our obligations to family come before our obligations to distant humans; our obligations to distant humans come before our environmental obligations; and critically for our concerns, our obligations to domestic animals come before those we have to wild ones.[22] In other words, he claims that highest priority should be granted to the community we have the most intimate relations with. Priority then descends as intimate relations do.

However, there are two important problems with Callicott's theory as outlined. First of all, if nearness is all that matters in terms of prioritising our obligations, then this can lead to some rather unwelcome implications. For example, it would mean that we can legitimately subordinate the interests of starving strangers in favour of lavishing completely useless extravangances on our nearest and dearest. While it is true that many of us in affluent societies do prioritise in precisely this way, that does not mean that this should be translated into legitimate public policy. In fact, many political theorists and campaigners for global justice are appalled by such partiality.[23] And interestingly, so is Callicott himself. For in response to these kinds of objection, Callicott has argued that partiality is not all that matters under his theory. He claims that when our obligations to two different communities conflict, not only should we consider the intimacy of the community, but also the strength of the interest. In this way, Callicott mandates that stronger interests take precedence over weaker ones. As such, he claims that because the interest faced by the threat of starvation is stronger than the interest in owning luxury goods, our obligations in fact lie with helping starving strangers.[24] But this concession is important, and Callicott himself underestimates its significance. For if under his theory the strength of an interest takes priority over the closeness of the community when working out our obligations, we have good reason to question whether it is really a partial and communitarian theory. In fact, it puts him in close proximity to the type of impartial theory advocated by thinkers like Singer, who judge interests simply on their strength and not by reference to who they belong to.[25]

Secondly, it is unclear that Callicott's theory has the implications for domesticated animals that he thinks it does. First of all, there is good reason to be sceptical of the kind of imagined contract between humans and domesticated animals that Callicott proposes. Could or would any domesticated animal really sign up to a life of a few months' free-range living followed by death by slaughter? Often, the benefits to animals of pastoral care under traditional farming methods can be overstated. Furthermore, we also have reason to question the extent of our intimate relations with domesticated animals in the mixed community. As John Hadley rightly points out, many of the animals in research laboratories and on factory farms are kept '...out of sight out of mind....'[26] Certainly, these animals may be domesticated, but most of us only interact with them once they are served up on our plates. As such, it does not seem that Callicott's communitarian theory offers domesticated animals the sort of strong protections that he desires.

In summary then, I believe that we have reason to be wary of the kind of partiality implicit in any kind of communitarian theory of justice for animals. Not only does it lead us into dangerous ground concerning the prioritisation of humans, and the prioritisation of particular humans, but it also rests on our rather arbitrary and conflicting feelings of concern for others.

Defining the community: Multiculturalism and animals

We have seen from the discussion so far that communities are not monolithic. They are not composed of individuals with identical sets of beliefs, values and goals. This makes delineating the common good of a society difficult. However, it also raises the question of *whose* common good it is that should be promoted. And this is the fourth problem facing the communitarian proposal for justice for animals outlined above. After all, in the discussion so far we have been assuming that the community whose values and norms are to be promoted is the state, and in our example, the UK state. But clearly states are made up of many different communities, including nations, religious groups, cultural groups, ethnic minorities and so on. Moreover, often these sub-state communities have values and practices that differ from those of the wider state. Indeed, sometimes members of such communities have stronger ties to the norms of their sub-state community than they do to those of the state. And clearly this is true regarding practices that relate to animals. Given this diversity, should the communitarian urge for the promotion of those values of the wider state, or of those of sub-state communities?

In practice, many governments attempt to promote the values of both. For example, some states have general animal welfare laws, but then grant exemptions from those standards to particular sub-state communities. To illustrate, consider two examples: religious slaughter and hunting. Taking religious slaughter first, in the UK, as in many other states, it is a legal requirement to stun an animal prior to it being slaughtered for meat. The purpose of this law is to ensure that all animals are rendered unconscious before they are killed so that they suffer less from the killing itself. This, it can be argued, promotes the value of avoiding the infliction of unnecessary suffering. However, the slaughter of animals used by the Jewish and Muslim communities is exempt from this law. This is on the grounds that the religious ordinances of these communities demand that in order for meat to be considered *kosher* or *halal* it has to come from an animal that is alive, conscious and in good health before it is killed. The traditional basis for such ordinances is obvious enough: clearly, it was to prevent people from becoming ill through eating the meat of diseased animals. However, the modern justification for such practices now simply rests on religious custom. But this custom causes suffering. In fact, the Farm Animal Welfare Council (FAWC), which advises the UK government on a range of animal welfare matters, has repeatedly urged governments to repeal the exemption on the grounds that it causes animals to suffer unnecessarily. Nevertheless, the UK continues to allow this exemption to these groups on the basis of freedom of religion.[27] Similar exemptions are in place in the USA, France, Australia, Germany and elsewhere. However, stunning before slaughter is a requirement for all communities in Sweden and Norway.

Sub-state communities are also sometimes granted exemptions from general animal welfare standards in relation to hunting. For example, indigenous people in many states are often granted exemptions from general legislation which outlaws the hunting of particular animals in particular places. Perhaps the most controversial example of such aboriginal hunting relates to whaling. The International Whaling Commission currently permits the aboriginal hunting of whales by several native peoples in Greenland, St Vincent and the Grenadines, the USA and Russia. Quite simply, these groups are given express permission to catch a certain number of whales per year, while members of their wider community are explicitly forbidden from hunting whales. As with religious slaughter, it is apparent that the aim of such permissions is to respect and uphold the values and practices of sub-state communities.

What are we to make of such exemptions? Well, it might be said that exemptions such as these are perfectly in accordance with communitarian theory. After all, communitarians seek to promote the shared values and goals of communities. Given that states contain a diversity of communities, it can be argued that it is only proper that the state seeks to promote the values and practices of the numerous communities that exist within it. As such, perhaps states should be applauded for adopting these 'multiculturalist' policies. It can be argued that by setting general standards for wider society, and then granting exemptions from those standards to particular sub-state communities, states successfully uphold the values and practices of *all* their communities, thus showing respect to *all* of their citizens. However, such multiculturalist policies are extremely controversial. Indeed, the use of such exemptions has been objected to on two important grounds.

First of all, we must remember that these exemptions are bad news for animals themselves. State legislation which demands that animals are stunned prior to slaughter, and which demands that individuals refrain from hunting whales, exists to protect animals. Such laws are not based on mere whimsy or preference, but on the reasonable ground that during the slaughter of animals for their meat, the least painful methods available should be used; and on the equally reasonable ground that killing animals as rare and as intelligent as whales is ordinarily a serious wrong. The award of exemptions from this legislation results in greater suffering for animals than if the legislation had universal application. Some communitarian thinkers, of course, might simply accept this. This increased animal suffering, they might argue, is just the price that we must pay in order to pay due respect to the norms of different communities. In the name of diversity and respect for different ways of life, they might argue, we should allow these communities to follow their traditional customs. But I wonder how far such thinkers would be prepared to go in applying such reasoning. Must we pay *any* price for the sake of respecting the norms of communities? Should we also grant groups exemptions from general standards that protect human beings? For example, some communities have norms which sanction quite brutal punishments for the transgression of moral and religious codes. These range from the payback system in Australian aboriginal law which can result in a spear through the leg, to the practice of 'honour killing' where individuals are killed for supposedly bringing shame on their family. Must communities which wish to practise such punishments be exempted from laws relating to assault and murder in the name of respect for diversity? Few scholars have endorsed such a position.

However, one thinker who gets close is Chandran Kukathas. Ingeniously, Kukathas eschews arguments based on diversity and respect for tradition, and actually uses liberal resources to argue for quite radical communitarian and multiculturalist policies. Kukathas argues that the human interest in freedom of association is of such importance that individuals should be permitted to live in communities with as little interference from the state in their lives and practices as possible. Talking of such communities, Kukathas writes:

> ...there would in such a society be (the possibility of) communities which bring up children unschooled and illiterate; which enforce arranged marriages; which deny conventional medical care to their members (including children); and which inflict cruel and 'unusual' punishment.[28]

This is extremely controversial stuff. Many, for example, would deny that the interest in free association is so important that it justifies the kinds of suffering and harms that these practices inflict. For it can be argued that while free association is one important interest for humans, it is by no means absolute, and certainly does not override other important interests, such as in being free from pain, fear and the threat of being killed. Moreover, many would question whether members of communities really have 'freely associated'. Rather, it might be argued that members have simply been born and socialised into their communities through no conscious choice of their own. And if that is the case, on what grounds can such cruel punishments be inflicted upon them? Moreover, and crucially for our purposes, it is clear that *animals* have not chosen to freely associate. Cows, chickens and sheep have not chosen to join communities which demand that animals be conscious when they are slaughtered; and whales have not chosen to join communities which demand that whaling is integral to their way of life. In which case, it is highly doubtful that such animals should be made to suffer from the practices of these communities.

The second objection that is often made against multiculturalist policies, such as those which exempt communities from animal welfare legislation, is that it is unclear who they apply to. Who counts as a community, and who does not? For example, it is evident that there are individuals and groups who would like to hunt whales because they enjoy the taste of their meat, or even because they want to make money from such hunting. It is also probably true that some smallholdings (and celebrity chefs) would prefer to slaughter animals themselves in their own backyard, without the trouble of taking them to registered slaughterhouses which have to comply with animal welfare legislation. Why are these individuals and groups not permitted exemptions when native peoples and religious groups are?

Taking the example of whaling first, perhaps there is a relevant difference here in terms of history. Native peoples have hunted whales in a sustainable way for generations, whereas other groups wanting to whale have not. This heritage, it might be argued, makes the exemption justified only in the case of native people. But does a practice's legitimacy really turn on how long it has been carried out? As discussed above, badger- and bear-baiting used to be important parts of British culture, as indeed was public flogging and execution. But now such practices are rightly considered cruel and barbaric. Perhaps then, the issue is less about the age of a custom, but its centrality to a community's way of life. For example, it might be claimed that many native communities are *whaling people* – whaling is so central to their culture that to prevent it is to effectively destroy that culture. But is that right? Are cultures really so bound up with particular activities like whaling that they will be utterly destroyed if they are prevented from continuing the practice? It certainly is not always the case. For example, it has been pointed out that when the Makah tribe in the USA asserted their 'cultural right' to kill whales in 1999 they had survived as a people despite having not killed a whale for over 70 years.[29]

But maybe things are clearer when it comes to religion. For perhaps exemptions from pre-stunning before slaughter are justified on the grounds of freedom of religion. Since there are no freedom of religion issues in the case of those individuals who simply want the convenience of backyard slaughter, on the other hand, perhaps an exemption is not justified. The problem with this argument, however, is that a universal standard requiring stunning before slaughter does not in fact interfere with religious freedom at all. Jews and Muslims are obliged to eat only that meat which is kosher or halal; in other words, they are obliged to eat only the meat of animals which were killed via a single cut, administered when the animal was conscious and carried out by a properly religiously authorised slaughterer. However, neither Jews nor Muslims are obliged to eat meat per se. Their religious ordinances do not forbid vegetarianism. As such, if pre-stunning were applied universally in the UK, meaning the end of kosher and halal meat, Jews and Muslims could still practise their religion freely. They would simply have to be vegetarian Jews and Muslims.[30] However, I suppose the more important point is that *even if* Jews and Muslims were obliged to eat meat, does that necessarily mean that they must be exempted from general animal welfare legislation? Does the fact that it is religiously based provide a justification for certain practices, a justification which reasons like age, diversity, freedom of association and so on cannot?

It is important once again to bear in mind that few would argue that religiosity warrants an 'anything goes' policy in relation to community practices.[31] Most thinkers rightly consider those acts which violate the basic rights of individuals to be unjustified, and do not warrant exemptions from general standards that protect such rights. That is why most consider practices such as polygamy, human sacrifice, burning at the stake, the drowning of witches and so on to be unjustified, even if based on religious ordinances. The relevant question then, is this: if human beings should be protected from those religious practices that violate their fundamental interests and rights, should not animals be similarly protected?

Now, at this point the advocate of multiculturalist policies might argue that there are double standards at play here. After all, even if it is conceded that practices such as religious slaughter and whaling cause animals to suffer, it might be pointed out that so too do the practices of intensive farming. In fact, it can justifiably be argued that the practices of intensive farming cause considerably greater suffering to the animals involved, and certainly involve greater numbers. On this basis then, Jews, Muslims and native peoples can argue that it is discriminatory to outlaw their customs, when these cruel actions are lawfully allowed to continue. This argument is indeed compelling and, in my view, more compelling than arguments based on diversity, free association, religiosity and so on. However, two things need to be pointed out in response. First of all, it cannot be claimed that applying animal welfare standards to all communities is *discriminatory*. The universal application of laws precisely does not discriminate. If the requirement for stunning and the ban on whaling were to be applied universally, this would not discriminate against these communities, it would merely impose burdens on them. It might be replied that the legislation imposes *unequal* burdens on those groups. But again, and rather obviously, legislation imposes unequal burdens on a whole range of individuals and groups: murderers are not allowed to murder, rapists are not allowed to rape, robbers are not allowed to steal and so on. Communities who want exemptions from general animal welfare standards thus need to show why the burdens they face are so important that they should be allowed exemptions, while others should not. And as we have seen above, arguments which invoke such things as tradition and religiosity are often on rather shaky ground. Secondly, Paula Casal has argued that while it may be true that it is hypocritical to outlaw these community practices when factory farms are allowed to indulge in similar forms of cruelty, two wrongs do not make a right. Casal argues that if states were unable to outlaw cruel practices until all similarly cruel practices were also outlawed, all attempts at incremental change would be thwarted.[32] After all, isn't it better that *some* suffering is avoided rather than none at all; and it isn't it better that we do *some* good, rather than none at all. Casal gives an interesting example to make her point:

> ... the fact that many murderous dictators have escaped the punishment they deserve is no good reason not to try other dictators whenever feasible – let alone to create legal exemptions for other individuals engaging in comparable activities.[33]

Of course, while Casal is quite right in arguing that in effecting incremental change, states do need to start somewhere, she must also recognise that in order for change to be truly incremental, states do need to go on and effect change elsewhere. And unfortunately, it is not evident that states are targeting the cruelty of factory farms. Of course, it is true that animal welfare legislation in many states is gradually evolving. But it must also be pointed out that the speed of that evolution is glacial. Billions of animals across the globe suffer terribly from intensive rearing systems and the political will to end that suffering appears to be weak. As such, perhaps sub-state communities have a point when they say that it is unfair to target their practices alone. Of course, that in no way justifies their own practices which are cruel to animals, but it does require reflection on the part of states about the consistency of their policies.

To sum up this long section, four points are worth re-emphasising. Firstly, any attempt to promote the shared values and norms of a community raises the question of *whose* values and norms are to be promoted. For as we have seen, states contain a number of different communities. Secondly, this is important in the case of animals because often states contain communities which have quite different attitudes and practices relating to animals when compared to those of wider society. Thirdly, one option for communitarian thinkers is to advocate 'multiculturalist' policies which allow communities to be exempt from general animal welfare standards. This allows for the goals and values of a range of communities to be respected and promoted. Finally, however, such policies are extremely controversial and have been objected to on the grounds that such practices cause real harm to individual animals, and because it is often unclear just which types of group warrant such exemptions and why.

Conclusion

This chapter has shown how communitarianism offers a political theory which is distinctive to utilitarianism and liberalism in its promotion of a politics of the common good. Unlike utilitarian theory, communitarians argue that the common good derives from the shared norms of particular societies. Unlike liberal theory, communitarians argue that the state can and should promote its common good, even when that interferes with the rights and liberties of individuals. As such, communitarian theory has the potential to put forward a theory of justice for animals that awards them strong protection. For instance, in societies which care

92 *An Introduction to Animals and Political Theory*

greatly about the interests of animals, communitarians can argue that the state should pursue strong policies of animal protection, and should pursue them even when they restrict the rights and liberties of individuals. In this chapter, I argued that the UK might be one society where the state should pursue such policies. However, we have seen that there are a number of problems with such a proposal. In the first place, a communitarian theory of this sort only provides protection to those animals that happen to reside within societies that value them highly. While some thinkers might regard this as a virtue of the theory, it is unclear why an animal – or indeed a human being – should have to suffer terribly simply because of where they happen to reside. Secondly, most societies contain a plurality of attitudes to animals, making attempts to ascertain definitive shared norms in relation to animals extremely difficult. For example, is there really an authentically British attitude to animals? Thirdly, because animal protection is based on how societies feel about animals, those animals that are unpopular in society, but who nevertheless have lives which can go well or badly for themselves, can be left vulnerable. While it can be argued that some degree of ethical partiality is only common sense, extending that partiality to the policy-making of states is much more questionable. And finally, any attempt to promote the common good must always define *whose* common good it is that should be promoted. As we have seen, most states contain a number of communities who often have quite different attitudes and practices relating to animals. If the social values of all communities are to be promoted, this can lead to the sanctioning of practices that cause animals significant levels of suffering.

Perhaps then it might be preferable to adopt a politics of the common good which is less particularistic. That is to say, rather than relying on the particular norms and values of a society, perhaps a theory of justice should just boldly state what is good for the collective, and promote that good, even at the expense of the rights and liberties of individuals. In a sense, this is what Marxist theories of justice attempt to do, and we examine such theories, and their consequences for animals in the next chapter.

6
Marxism and Animals

Marxist theory certainly shares something with communitarianism: it places high value on the communal life of human beings, and sees the ultimate human existence as one based on collective activity. Perhaps surprisingly, however, Marxism also shares something with liberalism. Like liberalism, Marxism is suspicious of state power and furthermore shares liberalism's concern with the *liberation* of individuals. How can this seeming paradox be explained? How can Marxism both prioritise the good of the collective and yet also be concerned with the emancipation of the individual?

To answer this, it is necessary to understand the peculiarity of Marx's own writings and the way that his theory differs from the other political theories that we have examined in the book so far. To understand this, we also need to move beyond some of the commonly held assumptions about Marx and his theory. This, of course, is a tricky task given the enormous influence that Marx's writings have had on the actual policies and institutions of many states. For example, it would be easy to characterise Marx as the proponent of a particular type of normative political theory which is opposed to capitalism, which champions the cause of the working class and which advocates collective ownership of the means of production. But while there is certainly an ethical dimension to much of Marx's work, and indeed to his political activity, it would be a mistake to regard his work as constituting a normative political theory in the traditional sense. For Marxism is not a theory about the principles and norms which should guide the institutions, laws and practices of the state. Rather, Marxism purports to be a scientific theory, *explaining* and *predicting* fundamental aspects of contemporary political and social life. Moreover, this scientific theory does not advocate an ideal set of political relations between state and citizens, but in fact predicts the *end* of politics and the end of the state. The scientific theory which Marx uses to make these explanatory and predictive claims is 'historical materialism', and it is through engaging with historical materialism that we can come to understand Marxism's prizing of the collective and the liberation of the individual.

The essential idea of historical materialism is that the history of humanity can be explained and understood by reference to the rise and fall of different modes of economic production. In all periods of history, human beings have been confronted by conditions of scarcity. As such, they have joined together in particular 'modes of production' to tackle that scarcity: slavery, feudalism, capitalism and eventually communism. Quite obviously, these so-called modes of production have different 'relations of production' – that is, they stipulate who has control over what in different ways. Furthermore, the political and legal institutions of a particular time in history – the 'superstructure' – arise to validate and sanction those particular relations. Hence we see Ancient societies possessing established norms and institutions legitimising slavery, feudal societies possessing norms and institutions legitimising hierarchy and loyalty, and capitalist societies possessing norms and institutions legitimising private property rights and individual liberty. Historical materialism claims that at the beginning of each mode of production, the system works well. The relations of production are validated and stabilised by the superstructure, which allows for sustained growth. However, the theory also claims that eventually the relations of production act as an impediment to growth. This clash between the relations and the growth of productive forces can only be overcome through revolutionary social change, and thus a new period of history. For example, it is possible to contend that with the advent of industrialisation, the strict concentrated and hierarchical ownership of land under feudalism was an impediment to growth. The only way to overcome this obstacle, according to historical materialism, was to overcome those social relations through bourgeois revolution. Now, the important point Marx wants to make is that what is true of feudalism is also true of capitalism. Capitalistic relations of production will also come to impede growth, particularly as technology advances. As such, these capitalist relations will need to be overturned, but this time by the proletariat, thus heralding communism. Importantly, however, communism will not just be another new mode of production. Rather, Marx claims that communism will be the *final* period of history: technological advance will overcome scarcity, the need for oppressive relations of power between individuals will thus cease, politics will end, the state will whither away and human beings will be truly liberated.[1] As such then, we arrive at the answer to our initial question. How does Marxism reconcile valuing both the common good and the liberation of the individual? By arguing that the individual is only liberated when human beings are collectively liberated under communism.

But while all of this is interesting in and of itself, what does it have to do with the treatment of animals? What does Marxism say about how political communities should govern their relations with animals? Well, in the first place, the obvious answer is 'nothing at all'. As a scientific theory, it can be argued, Marxism tells us nothing about what political communities should or should not be doing, let alone in relation to animals. Any attempts to answer such moral questions, it might be claimed, can only ever reflect the dominant relations of production. However, the analysis of Marxism and the treatment of animals cannot be left there. After all, Marxism cannot simply be reduced to this 'pure' interpretation of it. For one, it would of course be quite wrong to claim that Marx's own writings were totally devoid of normative claims about the governing of political communities. It is true that he regarded contemporary morality to be part of the bourgeois superstructure. But he also had a desire to be rid of that superstructure and its corresponding morality. His prediction that humans would be liberated under communism was also a *desire* that they be so liberated. This desire inescapably reflects a normative position.[2] Secondly, various forms of Marxist thought do have something explicit to say about the way in which political communities should be governed, many of which are relevant to our enterprise. For example, some of these theories are fiercely antagonistic to calls for greater protection of animals, while others have used Marxist resources in an attempt to propose quite robust forms of justice for animals.

As such, this chapter offers an overview and analysis of this relationship between Marxism and the treatment of animals by political communities. First of all, it looks at ways in which Marxism can be used to withhold justice from animals. This is done by first examining the qualitative differences between humans and animals that Marx proposed; and then by evaluating the charge that promoting the cause of animals impedes the achievement of proper socialist aims. Secondly, the chapter examines some of the Marxist resources that theorists have used in attempts to extend justice to animals. In this second half of the chapter, the issues of capitalist exploitation, using 'needs' as a distributive principle, and the question of effective political strategy are all examined in turn.

Discontinuities between humans and animals in historical materialism

All theories of justice which advocate strong protection for animals emphasise the commonality and continuity between humans and non-human animals. Whether it be the shared experience of life emphasised by Theophrastus, the shared capacity for suffering emphasised by Porphyry, Bentham and Singer, the shared capacity for reason emphasised by Plutarch or the shared capacity for intentional action as emphasised by Regan, this continuity between humans and animals serves as a central argument for the extension of justice to non-human animals. For all these thinkers, there is a morally relevant continuity between humans and animals. In stark contrast, in Marx's writings few commonalities between humans and other animals are presented. Instead, we see Marx regularly emphasising the qualitative *difference* between humans and animals. Of course this point should not be overstated; after all, Marx is fully aware that human beings are animals. Indeed, Marx's materialism certainly prevents him from regarding humans as divine, or in some special relationship with God. Humans, for Marx, are natural beings like everything else, and share many of their basic needs with animals.[3] Nevertheless, this human-animal continuity in Marx's writing does not extend far. Much more prevalent throughout Marx's work is the *special* status that he gives to human beings. Human beings may well be animals, but for Marx they are a very special sort of animal.

To explain, it is worth returning to the theory of historical materialism briefly outlined at the outset of this chapter. Recall that Marx's account of history is resolutely anthropocentric. World history is a history of humanity, a history which culminates in the emancipation of human beings. As we have seen, this emancipation involves humans being set free from scarcity, and in turn, the oppressive relations that flow from the conditions of scarcity. But that emancipation also entails humans being set free from nature, and from their animality. Emancipation is what allows humans to escape their natural predicament to become truly human. To explain what this means, it is necessary to delve into three concepts which Marx elaborates upon in his *Economic and Philosophic Manuscripts*: productive labour, alienation and species-being. For Marx, the essential human activity is productive labour. Intentionally labouring on the world, transforming the world and producing objects all define what it is to lead a truly human life. This free, conscious and spontaneous productive labour is what Marx calls humanity's 'species-being'. However, Marx goes on to claim that when an individual is forced to labour for someone else, as they are under wage-labour capitalism, then that individual is alienated. She is alienated from the product of her labour, in virtue of the fact that those products are handed over to someone else. She is alienated from the process of production, in virtue of the fact that she is forced to sell her labour in order to subsist on the wage she receives in return. She is alienated from her fellow workers, in the sense that everyone is judged by the relations they find themselves in as a worker. And finally, she is alienated from her essential human activity, from her own species-being, in the sense that her productive labour is not free and is not spontaneous.[4] Most importantly for our concerns, Marx claimed that this alienation reduces workers to mere animals:

> The result, therefore, is that man (the worker) feels that he is acting freely only in his animal functions – eating, drinking, and procreating, or at most in his shelter and finery – while in his human functions he feels only like an animal. The animalistic becomes the human and the human the animalistic
>
> To be sure, eating, drinking, and procreation are genuine human functions. In abstraction, however, and separated from the remaining sphere of human activities and turned into final and sole ends, they are animal functions.[5]

In other words, for Marx human beings cannot be truly human when they are made to labour under capitalism. Wage labour turns workers into animals, whereby they labour solely to satisfy their most basic needs. It is only through overthrowing these social relations that workers can be truly free, according to Marx. It is only when capitalism is abolished that human beings can be truly human. For Marx claims that under communism, scarcity will be no more, oppressive relations will be overthrown, and human beings will be able to produce freely and spontaneously for their own purposes:

> Not until this stage is reached will self-activity coincide with material life, will individuals become complete individuals. Only then will the shedding of all natural limitations be accomplished. The transformation of labour into self-activity corresponds to the transformation of the previous restricted interaction into the interaction of individuals as such.[6]

Of course, it might be objected at this stage that Marx's separation of human beings from animals is fallacious. After all, animals also labour. Animals also transform nature for their ends. Furthermore, under capitalism, animals are also made to labour for others. Why then cannot this theory of emancipation also apply to animals? Of course, Marx is fully aware that animals also labour. However, he regards animals' production to be of a different and lower order than that of human beings:

> To be sure animals also produce. They build themselves nests, dwelling places, like the bees, beavers, ants etc. But the animal produces only what is immediately necessary for itself or its young. It produces in a one-sided way while man produces universally. The animal produces under the domination of immediate physical need while man produces free of physical need and only genuinely so in freedom from such need. The animal only produces itself while man reproduces the whole of nature. The animal's product belongs immediately to its physical body while man is free when he confronts his product. The animal builds only according to the standard and need of the species to which it belongs while man knows how to produce according to the standard of any species and at all times knows how to apply an intrinsic standard to the object. Thus man creates also according to the laws of beauty.
>
> In the treatment of the objective world, therefore, man proves himself to be genuinely a *species-being*.[7]

In other words, for Marx, animals and humans are qualitatively different. Marx views historical processes solely in terms of the realisation of *human* liberation. The unfolding of history leads to a point where *humans* will eventually be able to labour freely and universally. Animals are of a different and lower order under this theory: they cannot labour freely and universally; they cannot develop or progress as a species as a whole; and they are thus not part of the unfolding of history. As such, we really have no reason or need to worry about animals at all. Historical materialism regards history as the development of *humanity*. It is thus the job of theorists to inform that process by pointing out the futility of investigating questions about how political communities should govern themselves, including questions about our proper treatment of animals. It is not, then, that Marx would have been opposed to extending justice to animals, it is more that he would have thought the very idea of it to be entirely irrelevant.

What are we to make of such an argument? In the remainder of this section, I will outline three important arguments which can be made against it: that its teleology is implausible, that its anthropocentrism is out of keeping with its materialism, and that its distinction between animals and humans is either false or morally irrelevant. In the first place, it is important to note that to most modern readers, the teleological nature of historical materialism is quite remarkable. The idea that history is unfolding towards some final and ultimate point where scarcity will be overcome, conflict will be no more, and politics and the state will whither away is simply too much for many to swallow. It is particularly hard to swallow in light of what we have learnt about the consequences of human development and technological progress for both the environment and our own well-being. Environmentalists have taught us that there are limits to human development: that technological advance will not be able to cope with unlimited growth in population size, resource use and environmental destruction. If there is any truth to this position, and if we are not necessarily headed towards some blissful future where abundance and communal harmony will reign supreme, perhaps then it is necessary to have some principles and norms by which to manage the institutions, laws and practices of politics. Perhaps too, it is necessary to have some such policies that relate to animals.

The second objection that can be made against this theory relates to its anthropocentrism. Marx purports to recognise that human beings are mere animals, thus rejecting any idea that they possess immortal souls granted to us by divine powers. Given his materialism, Marx's position here makes a good deal of sense. But in spite of his materialism, the reverence Marx gives to the human species is breathtaking. Humans are the sole subjects of history, and they are the sole makers of history. Furthermore, for Marx, human emancipation signals the end of history. As such, historical materialism is resolutely and remarkably anthropocentric. Marx, of course, was fully aware that the Earth was not the centre of the universe, and was also aware of Darwin's discovery that all animals – including humans – share common ancestry. Indeed, Marx fully accepted that human beings are part of nature. But these facts are difficult to discern when reading his theory of historical materialism. Indeed, historical materialism seems to grant human beings almost divine powers and divine status. However, if we take the view that human beings are part of nature more seriously, a more humble vision of humanity should probably emerge. That more humble vision would surely recognise that the history of the world is not all about, for or controlled by human beings. Moreover, that more humble vision would surely also accommodate the interests and needs of non-human animals, rather than viewing them as an irrelevance.

Thirdly, the great division between humans and animals that Marx erects in his discussion of productive labour can be questioned on a number of levels. Marx claims that human labour differs from animal labour in three ways: firstly, humans produce beyond physical need, but animals do not; secondly, humans can produce as a species, while animals can only produce as individuals; and finally, humans labour according to a range of standards, but animals only produce according to fixed standards for their species. But we can question both the accuracy and moral relevance of all of these alleged differences. First of all, it is not true that all humans produce beyond physical need, and that all animals do not. This is a gross generalisation that is undermined by some simple facts. For example, many human beings cannot produce at all due to age, illness or disability.[8] Many others cannot produce beyond physical need because of poverty, environmental catastrophe or other circumstances. And looking at the animal kingdom, it is evident that many species of animal can and do produce beyond physical need. For example, many predator animals, such as wolves or lions, will often kill more than they and their kin or pack physically need. The remnants of such kills, after all, provide sustenance for various scavenger animals that live in the same environments. Secondly, when Marx says that human beings produce 'as a species', once again he neglects the fact that significant parts of the human population are not capable of any productive labour whatsoever.[9] Furthermore, while it is true that human beings do have a collective potential to develop as a whole, it is not at all clear that these developments are always morally welcome, as the advent of nuclear weapons and environmental destruction is evidence of.[10] Indeed, it is certainly dubious whether human beings have the potential to develop forms of global collective human labour that would not require the kind of coercive political power that Marx believes will whither away under communism.[11] Finally, it is not true that animals can labour only according to a single fixed standard. Animals can adapt to new environments and labour in successful and innovative ways.[12] The success of many so-called 'non-native species' is clear evidence of this fact. And while it is true that human beings may be able to produce according to other and different standards, such as those defined by beauty or human culture, it is unclear what moral significance those capacities hold. It is just as true to point out that animals are able to produce in ways that we are incapable of, such as by using echolocation, spinning webs, using flight and so on. Does this fact then mean that such animal capacities are of greater significance than human ones? Clearly there are differences in the ways humans and other animals produce, but we have to question in what way those differences are morally relevant.

It may be useful to briefly summarise this long section. For Marx, humans and animals are qualitatively different. Human beings are the subjects of historical materialism, and it is human emancipation alone that signals the end of history under communism. This makes the entire idea of justice for animals entirely irrelevant for Marx. How political communities organise themselves and treat their animals is beside the point. History is about the unfolding of *human* relations. But as we have seen, these claims can be questioned on a number of levels. Firstly, we obviously have to question whether history does have the teleological kind of nature that historical materialism suggests. Secondly, we have to wonder whether the fiercely anthropocentric nature of historical materialism is at odds with Marx's own materialist position. And finally, we must note that many of the factual claims that Marx makes in respect of the qualitative differences between humans and animals are either false, or morally irrelevant. Perhaps then, theorists with Marxist sympathies should drop some of the anthropocentrism of historical materialism, recognise the continuities between humans and animals, and take the idea of justice for animals more seriously. In the next section, we see why a number of scholars on the left would be strongly opposed to such a proposal.

Animals rights as bourgeois morality

The relationship between animal protection and left-wing politics has always been ambivalent. On the one hand, many of the most important campaigners of justice for animals have also been socialists, and the two movements do have similar and close historical roots. On the other hand, however, many others on the left have viewed the animal protection movement with outright hostility. Such opponents have considered animal rights at best as a reflection of elite and middle-class interests, and at worst as an obstacle to the achievement of serious socialist aims relating to the emancipation of human beings. Before examining in detail these leftist critiques of animal rights, let us first consider some of the historical associations between animal protection and left-wing politics.

The link between socialists and animal campaigners was forged largely in the social reform movement of Victorian Britain. At this time, a number of radical social reformers campaigned on a range of issues, including the abolition of slavery, the enfranchisement of the working class, the emancipation of women and the protection of animals. Many radicals of the time considered these issues to be linked, and believed that radical societal reform in relation to them all was both necessary and possible. For example, William Wilberforce, the anti-slavery campaigner was also a founding member of the Royal Society for the Prevention of Cruelty to Animals (RSPCA). Lord Shaftesbury, the author of the Factory Acts, was also a campaigner against cruelty to animals. The Humanitarian League, created by Henry Salt, counted George Bernard Shaw and Ramsey MacDonald amongst its members, and campaigned for benevolence to be shown to all sentient life, including animals.[13] The question of justice for animals was a live issue in Victorian Britain, especially in London where over a dozen vegetarian restaurants were operating successfully by 1886.[14] Importantly for our concerns, the issue of animal protection sat comfortably alongside the other important social reform issues of the time. Many of those Victorian social reformers were both socialists and campaigners for animal protection. And indeed, there is some evidence that this relationship between socialism and animal protection has carried forward to modern times. For example, Robert Garner's research on the UK and USA has shown that in the 1990s animal protection policies were mainly supported by centre-left parties, and mainly by individuals towards the left of those particular parties.[15]

However, it would be quite wrong to consider the relationship between socialism and justice for animals as a marriage made in heaven. Many on the left have been openly hostile to campaigns for better protection for animals, believing the issue to be frivolous, reflective of bourgeois morality, and an unnecessary and unwelcome obstacle to the serious business of ending the oppression and suffering of human beings. For example, many on the left view the Victorian social reformers described above not as progressive socialists, but as elite protectors of bourgeois interests. In this spirit, Howard L. Parsons writes as follows:

> Humane societies for the prevention of cruel and 'inhuman' treatment of animals were a natural accompaniment to the movements for social reform. Humane societies and conservation groups, however, tend to arise among the wealthy classes and higher-salaried or professional persons impelled by a variety of motives: a sense of 'ownership' and identity with one's country, a desire to protect one's own private holdings, humanistic idealism, an elitist fear of

popular or socialist control of resources, and a diversion and displacement of energy from the radical transformations demanded in
society. (Some Nazis were fond of animals and believed in the conser-
vation of nature.) Often the ruling and affluent classes expend great
energy and time on the protection of humanized animals rather than
on the welfare of brutalized children in home and factory or on adult
workers reduced to the level of animals. Their concern for animals is
a displacement of human concern, for their class position constricts
the scope of their expressed human concern.[16]

Importantly, many on the left do not restrict this charge to the Victorian
social reform movement. They also level it at current campaigns for
animal rights. The charge is that concern for animals is a bourgeois affec-
tation, which says more about human beings and their interests than it
does about the suffering of animals.[17]

In order to address the claim, it is necessary to break it down into
its constituent parts. After all, there seems to be a number of different charges being levelled against the animal protection movement in
these attacks, each of which must be opened up to scrutiny and evaluated. In the first place, there is the claim that concern for animals is
an elite pursuit conducted by the middle classes. The idea seems to be
that only the wealthy can afford to care about animals because the poor
have more important matters to concern themselves with. However, not
only is this claim patently false, but it is also incredibly patronising. It is
simply not true that animal protection is an issue for the wealthy only.
Firstly, some of the strongest movements for animal protection are based
in developing countries like India. Moreover, it can even be argued that
the concern for animals in such countries is in fact diminishing as the
level of wealth amongst the population increases. Secondly, in developed states, campaigners for animals obviously come from a huge range
of socioeconomic backgrounds, including from the poorest sections of
society. And this should not be surprising: the rich are obviously not
the only individuals capable of feeling compassion for the suffering of
animals.

However, let us for the sake of argument assume that the rich do
have a monopoly on compassion for animals. What ethical conclusions
should we draw from that hypothetical (and false) fact? Obviously, we
should draw none at all. An ethical position can and should be judged
independently of considerations of its source. For example, even if calls
for the abolition of slavery, the enfranchisement of the working class
and the emancipation of women came exclusively from the middle

classes (which they did not), that does not make such calls ethically flawed. And the same is true for animal protection. Even if it were true that all animal rights campaigners were wealthy (or even that some were Nazis!) that does not make their moral claims invalid as such. All normative claims should be judged on their own merits, irrespective of their source.

Another part of the charge against animal protection movements seems to be that the motivation of their members does not really derive from concern for animal interests. Instead the claim is that the motivation behind animal protection movements really derives from human interests, such as the protection of an elite position, or the imposition of a 'superior' moral code. However, seeing as we cannot enter the minds of campaigners for animals, such a charge is impossible to prove. Furthermore, the charge has more than a whiff of paranoia about it. No doubt there are some in the animal protection movement who are out to further their own interests: to show how clever and righteous they are, or even to gain fame and celebrity. But rather obviously, such individuals can be found in all movements, professions and walks of life. It is wrong to tar the animal protection movement as a whole with this brush, especially given the evident sincerity and selfless good work of so many in the movement. And it is also surely wrong to treat a social movement with suspicion just because it takes a moral stand. Social movements have certainly helped to transform society for the better in a number of ways on various issues, whether it is civil rights, equality for women or discrimination against the gay community. To judge each and every movement as suspicious for being elitist and self-interested is not only curmudgeonly, but also incredibly reactionary.

The final charge made by those on the left against the animal protection movement is that it somehow diverts attention and resources from the real and more pressing problems of the day. Parsons in the quotation above calls concern for animals a '... displacement of human concern', and goes on to state that Marx and Engels were interested in the starvation and disease of the human population, and not with '... displaced sympathy for the animals....'[18] This final element of the charge is probably heard the most often; the idea being that worrying about the suffering of animals is foolish and decadent given the suffering of so many human beings around the world. However, two responses can be made against this charge. First of all, we can question why we have to choose between caring for animals *or* for humans: for why should we not show concern for *both*? After all, it is surely wrong to regard moral compassion as a finite resource that can only be extended to X number of other individuals, or to X number of specific issues. Many individuals, including John Stuart Mill, Gandhi, Henry Salt and so on, have shown that compassion can easily extend across a number of domains and issues, including to animals, without contradiction. Clearly, we may have to make some tough choices when the vital interests of humans and animals clash. But in reality, how often do such clashes occur?[19] After all, the most extreme forms of animal suffering take place in intensive agriculture, and political communities can easily take action to prevent such suffering without sacrificing their own vital interests. Secondly, not only should we question the idea that we must choose between caring for humans *or* for animals, but we must also question whether concern for animals is decadent. For if the unnecessary suffering of sentient individuals counts at all, and I believe most people would agree that it does, then something like intensive agriculture must surely be one of the most urgent moral issues facing contemporary societies. This is quite simply because the numbers of animals involved in these processes are vast, running into the billions across the globe annually. And given that all humans can live perfectly flourishing and successful lives without such cheap animal products, that suffering can also be regarded as entirely unnecessary. Given these simple and uncontroversial facts, the issue of intensive animal agriculture certainly seems far from trivial.

Some theorists on the left have acknowledged the weight of the moral problem raised by the treatment of animals in modern societies. In fact, they have argued that Marxist resources can be used to address that problem, ameliorate the position of animals and extend justice to them. Some have even suggested that a properly reconstructed Marxism demands that justice be extended to animals in this way. In the remainder of this chapter, we will examine three attempts to further the cause of animals using Marxist resources.

Animals as an alienated and exploited group

The first way in which some theorists have drawn upon Marxist resources in order to extend justice to animals is through drawing parallels between the position of workers and animals under the capitalist mode of production. One claim has been that animals can and should be considered as an exploited class.[20] Katherine Perlo, for example, has argued that both workers and animals are exploited, in that neither receives the full value of their productive activity; instead, the capitalist extracts the surplus value from the work of both.[21] Since both animals and workers are exploited in this way, Perlo argues that Marxist theorists must be concerned with the plight of both.

In a similar vein, Barbara Noske has argued that just as human beings are alienated under capitalism, so too are animals. To make her point she considers each form of human alienation as proposed by Marx, and applies it to farm and laboratory animals. For example, Noske claims that just as human beings are alienated from the product of their labour, so too are animals: 'Animals are being alienated from their own products which consist of either their own offspring or (parts of) their own body.'[22] Indeed, by giving their own bodies and lives to the capitalist, the alienation from product that animals endure can surely be regarded as total. Noske also claims that animals are alienated from their productive activity under capitalism.[23] She points out that animals are usually forced to concentrate on just a single productive activity, such as fattening, at the expense of all of their other natural activities. Thirdly, Noske argues that animals are alienated from their fellow animals under capitalism by being removed from their natural social arrangements, and by being forced to endure conditions so cramped that no normal forms of social bonding can take place.[24] Indeed, the prevalence of such behaviours as tail-biting amongst pigs and pecking to death amongst poultry, is clear evidence that factory-farmed animals lead distorted social lives. Fourthly, Noske considers a form of alienation not mentioned by Marx, which is the alienation from nature. Here Noske claims that capitalist production removes animals from their natural ecosystems, from their natural stimuli and from their natural behaviour patterns. Finally, Noske argues that all of these forms of alienation collectively amount to animals being alienated from their species life.[25] While Noske does not elaborate on this point, it would be easy to relate this form of alienation to Martha Nussbaum's capabilities approach which was discussed in Chapter 3. You will recall that Nussbaum equates an animal's well-being with its ability to perform the valuable natural functionings of its species.[26] This certainly resonates with Marx's idea of species-being. Moreover, it could well be argued that animals in intensive forms of agriculture are alienated and harmed to the extent that these natural functionings are impeded: pigs are prevented from exercising such functionings as rooting; poultry are prevented from exercising such natural functionings as nesting and scratching; dairy cows are prevented from such natural functionings as bonding and caring for their calves and so on. Once again, it can be claimed that just as Marxists should be concerned with the alienation of workers under capitalism, so too should they be concerned with the similar forms of alienation suffered by farm and laboratory animals.

The parallels between workers under capitalism and animals under capitalist modes of intensive agriculture are certainly striking. And it is certainly true that if we take the Marxist concepts of exploitation and alienation seriously and employ them as useful tools of analysis for human beings, there seems to be no good reason why they should not be similarly applied to non-human animals. However, just because Marxist concepts and forms of analysis can be applied to animals that does not mean that Marxism and the extension of justice to animals are in perfect harmony. Perlo and Noske both claim that Marxists should be concerned with the exploitation and alienation of workers and animals. Presumably, the idea is that emancipation for both can only be achieved by overthrowing the oppressive relations of capitalism. However, one can question this causal relationship between capitalism and animal exploitation in two ways. Firstly, it is doubtful that capitalism actually *causes* animal exploitation.[27] To explain, it is obviously true that certain features of capitalist production, such as the search for profit and competition between food producers, have exacerbated the suffering of animals. Without doubt, the urge to produce huge quantities of meat with great efficiency has led directly to the forms of intensive animal agriculture we see today. Animals are bred, confined, fattened and slaughtered in conditions which are purposefully designed to produce more meat at lower cost, and thus to boost profits. These intensive forms of production provide increased profits for capitalists – and cheaper meat and dairy products for us – but at the expense of terrible animal suffering. As such, the claim that capitalist systems of production have helped to intensify animal suffering is pretty hard to refute. Nevertheless, it is much more difficult to hold that capitalism actually *causes* such suffering. For it is obviously true that other modes of production did, have and could exploit and alienate animals. Looking back through history, for example, it is certainly evident that other forms of production have both exploited and alienated animals. It is all too easy to review agriculture before factory farming with rose-tinted spectacles, and as embodying a system of idyllic and benevolent pastoral care. For it is worth reminding ourselves of the obvious fact that animals have suffered from being used in agriculture ever since they were first domesticated. Clearly, the numbers of animals involved and the forms of suffering may have been less intense than under modern forms of intensive agriculture. But still, Ancient and feudal societies certainly confined, kept and killed animals in ways which alienated them from their natural functionings, and in order to extract profit from them.

As well as questioning whether capitalism is the cause of exploitation, we must also consider whether capitalism is a necessary impediment to achieving justice for them. This is a large and difficult topic, and there is not the space here to delve into it in any real depth. But nevertheless it is at least worth remarking that capitalist relations might not be a necessary obstacle to achieving justice for animals. For example, it might be argued that there is nothing intrinsically wrong with extracting surplus value from the labour of domesticated animals, so long as those animals are not harmed in the process. In other words, it might be argued that there is nothing intrinsically wrong with, say, raising and keeping chickens to lay eggs to sell on for profit, or raising and keeping cows for their milk to sell on for profit. There is no wrong done in these scenarios, it might be claimed, so long as the animals involved are not made to suffer. Of course, such animal use would likely be considered exploitative by Perlo and Noske, and they would be correct under a strict Marxist use of the term. However, exploiting animals in ways which do not cause them to suffer can coherently be argued to be permissible. After all, if the chickens and cows are raised and kept in ways that enable them to lead enjoyable lives free from suffering, it is certainly *difficult* to identify any wrongdoing. Obviously, my point is not that current capitalist production methods treat animals in this way. Quite evidently they do not: they make animals suffer terribly. Rather, my point is that capitalist modes of production could conceivably be altered and changed in ways that are benign for animals. It is at least possible to envisage a capitalist society which both raises animals for profit, and yet which does not cause them harm. As such, capitalism might not be the impediment to justice for animals that some Marxists see it as.

In summary, while it is possible to view animals as an exploited and alienated group, that does not mean that Marxism is a perfect ally for the extension of justice to animals. Given that capitalism does not cause animal exploitation, and given that justice for animals might be achievable under capitalist relations, the relationship between the two movements might not be as close as some writers have suggested.

'To Each According to their Needs'

The second way in which thinkers have used Marxist resources in order to extend justice to animals is through use of the concept of needs. The famous socialist principle, 'from each according to his abilities, to each according to his needs', was endorsed by Marx in his article 'Critique of the Gotha Program':

> In a higher phase of communist society, after the enslaving subordination of the individual to the division of labour, and thereby also the antithesis between mental and physical labour, has vanished; after labour has become not only a means of life but life's prime want; after the productive forces have also increased with the all-round development of the individual, and all the springs of common wealth flow more abundantly – only then can the narrow horizon of bourgeois right be crossed in its entirety and society inscribe on its banners: from each according to his abilities, to each according to his needs![28]

Here Marx is putting forward the idea that in the advanced stages of communist society it will not be necessary to be overly concerned about how resources are distributed, one of the prime concerns of contemporary liberal political theorists. For in this stage, we will have transcended all need for bourgeois schemes of distributive justice. Abundance will allow resources to be produced in accordance with how individuals are able to produce them, and distributed in accordance with how individuals need them. Importantly for our concerns, a number of theorists have seized on this maxim and used it as a basis for extending justice to animals. For if the focus of our principle of justice is the needs of individuals, then since animals quite clearly possess needs, that principle can be extended to them quite straightforwardly. For example, David Sztybel has argued as follows:

> I argue that it is essentially speciesist to restrict this principle to human beings, and that its acceptance implies either animal rights or a substantive equivalent. Marxism may have to undergo a profound dialectical transformation in light of the implications of its own maxim.[29]

Ted Benton has also argued that the maxim can be read as a principle of distributive justice, and can be used to extend justice beyond the species boundary. In fact, Benton argues that a focus on animals' needs may actually be superior to a focus on rights in terms of achieving justice for them. For while many people have philosophical problems with attributing rights to animals, few have trouble recognising that they possess needs.[30]

However, there are three reasons to be wary about using this principle in order to extend justice to animals. The first reason to be wary is that it is unclear how central this principle is to Marxist thought. After all, Marx is certainly not the author of the maxim, and under some interpretations of the passage quoted above, considers it to be of little importance. For example, in the quoted passage Marx is not declaring that resources must and should be distributed strictly in accordance with this maxim. It is not a principle of justice like those of John Rawls, for example. Rather, he is saying that under the advanced stages of communism we do not even need to worry about the distribution of resources. Life will be so different under communism – individuals will be liberated, resources will be abundant – that we do not need to concern ourselves with distributive principles at all. As such, Marx is pointing out that each can have as much as they need. The maxim can thus be interpreted as a throw-away line, telling us not to worry so much about distributive principles.[31] If this interpretation is correct, then one wonders if this Marxist resource is really so useful for extending justice to animals.

Secondly, and related to the above point, the maxim is not intended for use in contemporary societies. Marx is quite clear on this in the passage above. This is a principle for the advanced stages of communist societies. Given that Marx did not mean it to regulate distribution amongst humans in contemporary societies, one has to question whether it should then be applied to animals in those same societies. To his credit, Sztybel does acknowledge this objection and tackles it head on: 'If we are working towards a communist society, then the communist principle remains our ideal, to be fulfilled insofar as possible'.[32] In other words, Sztybel recognises that this is not a principle that we can realise here and now in conditions of scarcity, but it is a goal that we can do our best to approximate as far as we possibly can. However, mining this ethical imperative from Marx's work is extremely controversial. There is little evidence from the texts to suggest that Marx would endorse the kind of 'approximation' that Sztybel suggests. Indeed, it is likely that he would regard it as entirely pointless. As noted above, the job of the theorist for Marx is not to come up with normative principles to regulate what political communities ought to do, but to direct attention to the futility of such efforts given the unfolding of history. Furthermore, since we have seen that Marx did not really regard this maxim as a normative principle of justice anyway, but instead used it to illuminate the remarkable transformation that communist society will bring about, it is highly unlikely that he would have endorsed the kind of approximation proposed by Sztybel.

Nevertheless, just because Marx himself may have rejected the use of the principle in the way endorsed by Sztybel, that does not mean that it is useless. For we are under no obligation to faithfully follow the strictures of Marxist thought. Perhaps then we should employ it as a useful means of extending justice to animals, irrespective of whether it conforms to Marx's own work and convictions. In this sense, perhaps we can use the principle as a *Marxian* resource, rather than a strict *Marxist* resource, in order to afford strong protections to animals. Unfortunately, however, there is a third reason to be wary about using the principle in this way. This third problem relates to the substance of the principle itself. For one, the principle is obviously incredibly vague, and thus of limited practical use. For example, what are the needs of animals? Sztybel argues that '... every being needs to be free from suffering, and needs to live, to exist at all'.[33] However, is being free from suffering really a need? Many would argue that needs relate to those goods necessary for *survival*, and not to such conscious states as pleasure and pain. As such, a farmer using intensive methods might well claim that he meets the basic needs of his animals, by keeping them alive and as free from disease as possible. In this way, many forms of intensive animal agriculture might be in accordance with this Marxian maxim, prompting us to question the level of protection that it affords to animals.

However, let us put such difficulties to one side for the moment. Even assuming we can and do know the needs of animals, there is a deeper problem with using this principle to extend justice to animals: it mandates that we extend justice well beyond mere sentient animals. After all, all living organisms have needs which must be satisfied in order to make them better examples of their kind. Using this communist maxim as a distributive principle thus seems to suggest that political communities should devote energy and resources to satisfying the needs of plants, viruses, bacteria and so on, as a matter of justice. While Benton is quite happy to accept this implication, and in fact sees it as a virtue of the theory, most thinkers will regard it as far too demanding, and completely impractical, given the sort of need-clashes between living organisms that are endemic in all societies and environments.[34] As such, and in order to avoid the principle being extended in this way, Sztybel claims that the principle only applies to the needs of sentient animals. The reason for this, according to Sztybel, is that only those needs attached to conscious experience are morally relevant: 'Things matter to animals but not, one could argue, to plants, and the latter are oblivious to their own survival and death.'[35] But this move by Sztybel is rather unsatisfactory. After all, his whole argument is premised on the injunction that we apply this socialist principle in a non-prejudicial way. He argues that just as animals have needs, so too do animals. It is needs which count for justice, according to the principle, and thus to restrict this principle of justice to humans would be speciesist. However, we later learn that needs are not in fact doing the job. The principle does not apply to all those with needs, according to Sztybel, but only to those who are sentient. The obvious reply must be that if it is sentience, and not needs, which matters for justice, it would surely be better to have a principle with sentience as its focus, rather than one focused on needs.

As we can see from the above, there are three very important reasons to be cautious about using the socialist principle, 'from each according his abilities, to each according to his needs', to extend justice to animals: firstly, we can question its centrality to Marxist thought; secondly, we can question its applicability to contemporary communities; and finally, a focus on needs can be both vague and overly demanding.

Achieving justice for animals politically

The final means by which Marxism has been used to extend justice to animals is as a resource for thinking about how to achieve that goal *politically*. What I mean by this is that Marxist thought has been used to inform the animal protection movement about how they might achieve change in the real world. The leading proponent of this view has been Ted Benton. Benton's key point is that a liberal rights-based strategy for achieving justice for animals will never prove successful. He argues that assigning formal rights to individuals is always pretty useless unless those individuals can *enjoy* those rights, and this is as true for animals as it is for human beings. Moreover, and crucially, Benton argues that under present capitalist relations, individuals – including animals – will never be secure in their rights: 'Social, economic and political relations of power and dependency, dominance and subordination may serve to render rights, even where formally acknowledged, more-or-less ineffective.'[36] For example, Benton argues that a liberal rights-based strategy is particularly ineffective in achieving real change for animals when confronted with the economic and political power of agribusiness and pharmaceutical companies. Attempting to *persuade* business and politicians that they should protect the basic interests of animals used in agriculture and laboratories is going to be pretty ineffective given the enormous economic gain that can be made out of violating those interests. As such, Benton claims that this liberal strategy for securing protection for animals must be complemented by, and even replaced by, a broader strategy aimed at effecting wide-ranging shifts in owner-ship and wealth.[37] In effect, Benton is claiming that in order to achieve justice for animals, it is necessary to instigate changes in the social and economic relations which perpetuate injustice towards them. In other words, animal campaigners need to pursue leftist goals in order to secure protection for animals.

This is certainly a fascinating argument, and in my view offers the most valuable form of alliance between Marxist theory and strong forms of animal protection. For there is certainly no doubt that campaigners for animal rights need to do more than simply get those rights written into the law. Having formal rights written in the law is one thing; having them protected so that they may be enjoyed is quite another. Neverthe-less, Benton's critique of the liberal political strategy is a little puzzling. Benton argues that rational persuasion regarding the need to protect ani-mals is insufficient, for in order to achieve justice for animals, we also need to effect social, political and economic change. Maybe he is right on this point. But even if he is, how are we to effect the kind of change that he describes? That is, how are we to effect such change when so few others are interested in it? After all, it is perfectly evident that very few people have a real desire to fight for radical societal change, and very few have an interest in achieving justice for animals. Given such apa-thy, surely the only tool at our disposal is to persuade them that they should be interested in achieving justice for animals. That is to say, the only way we can overcome such apathy is to expose unjustifiable animal suffering and to provide good arguments to explain why such suffering should concern political communities.

Benton might be right in his claim that achieving justice for animals requires quite radical transformation. However, that transformation is never going to occur when people have no interest in it. As such, the liberal strategy of persuading individuals and politicians of the need for animal rights and better animal protection laws seems not only nec-essary, but absolutely indispensable if the goals of the animal rights movements are to be realised politically.

Conclusion

Delineating the relationship between Marxism and justice for animals is difficult given Marxism's unique angle on the question of justice itself. For as we have seen, Marxism purports not to outline normative principles to guide the laws, institutions and practices of contemporary political communities; rather, it claims to explain and predict key aspects of our political and social life through the theory of historical materialism. According to Marx, history can be explained and analysed through the unfolding of different economic systems. Moreover, he claims that the emancipation of human beings will only be achieved under communism, the final mode of production, when abundance will free humans from the oppressive relations that have characterised previous epochs. The question of how contemporary political communities should treat animals is thus something of an irrelevance for Marxists. For one, no true justice can be achieved while we live under the oppressive conditions of capitalism; and secondly, the emancipation heralded by communism is resolutely anthropocentric. As we have seen, however, important aspects of historical materialism can be challenged: its teleology can be questioned, not least because of recent environmental concerns; and its anthropocentrism can be challenged, not least because of Marx's own materialism. Many Marxists have acknowledged these problems. That is not to say, however, that there has been a significant groundswell amongst Marxists to extend and promote justice for non-human animals. For in actual fact, many on the left have maintained that concern for animals is a form of decadent bourgeois morality, which diverts energy and attention away from tackling the exploitation and alienation engendered by capitalism. As we have seen, however, this kind of either/or approach to important social and political problems is deeply unsatisfactory. Surely, we have the resources and energy to be able to identify and tackle more than one injustice at a time. Indeed, others on the left have attempted to use Marxist theory to further the advancement of animals' interests: by focusing on their status as an exploited group; by including them in a needs-based distributive principle; or by using Marxist insights to further the cause of animal rights politically. While each of these strategies is intriguing, they do also face significant problems. After all, the relationship between capitalism and the exploitation of animals is not as clear-cut as many of these authors suggest. Also, there are important doubts about the centrality of needs-based distribution to Marxist thought, as well as problems about extending such a principle to other needy living organisms, such as plants. Finally, while Marxism does offer important insights concerning the necessity of fundamental societal and political transformations for animals to enjoy their rights, it is unclear how such changes can occur unless people are *persuaded* of their necessity.

7
Feminism and Animals

In the last chapter, we saw that irrespective of the theoretical links between Marxism and justice for animals, the two *social movements* related to them possess close historical connections. Many Victorian social reformists, such as George Bernard Shaw and Henry Salt, campaigned on behalf of both workers and animals. Interestingly, the historical links between animal protection and feminism are just as strong, if not stronger. Once again, those connected roots lie in Victorian Britain where social reformers campaigned on behalf of women *and* animals. For example, Frances Power Cobbe was an active campaigner for women's suffrage, and also co-founder of the Victoria Street Society for the Protection of Animals Liable to Vivisection. Anna Kingsford was one of the first women in Britain to qualify as a doctor, did so without vivisecting a single animal, and campaigned against vivisection for the rest of her life.[1] Furthermore, it is certainly possible to argue that such connections have carried forward to contemporary times, not least because women comprise the bulk of campaigners in the animal protection movement.

But in addition to the links between these respective social movements, a number of feminist scholars have also made important *theoretical* connections concerning the issues of justice for animals and justice for women. Such feminists have claimed that the oppression of animals and women are interrelated, and that the achievement of justice for animals and women are interdependent. Furthermore, it has been argued by these scholars that the achievement of such justice depends on a properly constructed feminist political theory. As we will see, this properly constructed feminist political theory not only offers a critique of the norms, policies and institutions of modern political communities, but also offers a powerful critique of much of political theory itself. This critique argues that political theory should shed itself of its valorisation of reason, and instead concentrate on the feelings of care we have for others. To explain, you will recall from Chapter 2 how Western political theory has been dominated by 'reason'. Not only has the possession of reason been considered by many political theorists as a necessary ingredient in determining who gets justice, but reason has also been the prime methodological tool for determining what justice is. As we have seen, this prizing of reason derived from the work of Plato and Aristotle, continued in the thought of the Stoics and Roman writers like Cicero, was repeated by medieval Christian writers such as Augustine and Aquinas, continued in Enlightenment thinkers such as Locke and Kant, and of course lives with us today in the work of the majority of mainstream Western political theorists, such as John Rawls. However, a number of important feminist writers have rejected this valorisation of reason. First of all, they have rejected it as a basis for determining who gets justice, pointing out that it has been used to exclude and oppress both women and animals. Secondly, they have rejected it as a method-ological tool for determining what justice is, arguing that it neglects the important and foundational role that sentiment and emotion play in our ethical decision-making. For so-called 'care-based' feminist theorists, reason should be replaced, or at least supplemented by, feelings of *care*. As such, they claim that justice should be *extended* to those we do and ought to care for; and justice can be *determined* by reflecting on our sen-timental and emotional attachments to others. This feminist care-based theory has important implications for animals: it provides a novel ratio-nale for extending justice to animals; and it offers significantly different ideas about what treating animals justly involves.

However, it is important to point out at this stage that not all femi-nist theories are care-based, and nor are all feminist theories concerned with the question of justice for animals. There is no single and defini-tive 'feminism', but instead an extremely diverse set of feminist theories. While these theories are united by their concern to end the oppression of women, how that oppression is defined and the means by which it can be eradicated is a subject of great debate amongst feminists. For example, liberal feminists are primarily concerned with securing equal rights and opportunities for women. More radical feminists, on the other hand, argue that those rights and opportunities are pretty useless while wider patriarchal cultural norms remain intact. Importantly, the bulk of feminist writers have not been particularly concerned with the question of justice for animals. Just as the majority of utilitarians, liber-als, communitarians and Marxists have largely ignored the animal issue, the same is true of most feminists. For most of these thinkers, how polit-ical communities should treat animals has not been considered worthy of much analysis. Nevertheless, care-based feminism is an exception. This particular strand of feminist thought has taken the issue of our obli-gations to animals very seriously, and has developed an important and distinctive means of addressing those obligations. As such, the remain-der of this chapter will focus on care-based theory by analysing three of its most important claims that relate to animals. First of all, it will evaluate the claim of some feminists that the oppressions of animals and women are interrelated, and thus that their liberations are interde-pendent. Secondly, it will analyse the view put forward by care-based theorists that traditional reason-based means of extending justice to animals are fundamentally flawed. Finally, it will explore and assess care-based theorists' own proposals for how political communities should govern their relations with animals.

The linked oppressions and liberations of animals and women

Several feminist scholars have argued that the oppressions of women and animals are interrelated, and thus that their liberations are inter-dependent. In other words, they argue that because the marginalisation and exploitation of women and animals derive from the same sources, liberating one group necessitates liberating the other. In fact, these fem-inist writers usually move beyond women and animals, making links between the oppressions of *all* marginalised groups:

> Women, people of color, queers, non-human animals are all thought to be lower in the hierarchy than white, heterosexual, able-bodied human men. The conceptual tools and institutional structures that maintain the status of these men are employed against women and animals. Oppression of any of these groups is thus linked, and if one is opposed to sexism, racism, heterosexism, etc. she should also oppose speciesism.[2]

Of course, in order to evaluate the claim that these different oppressions are linked, it is necessary to understand how they are linked. Just what are these 'conceptual tools and institutional structures' that maintain this hierarchy? From the feminist literature on this topic, it is possible to delineate four ideas which are claimed to link and maintain the oppres-sions of women and animals: the domination of nature, the valorisation of meat-eating, forms of language and objectification through property status. This section will briefly outline each in turn, before evaluating these links at the end of the section.

The idea that the domination of nature has led to the oppression of both women and animals stems from 'ecofeminist' thought. The fundamental tenet of ecofeminism is that the exploitation of women and the environment are inextricably linked through the patriarchal domination of nature. Care-based feminists have simply extended this idea to include the exploitation of animals. As Josephine Donovan writes:

> From the cultural feminist viewpoint, the domination of nature, rooted in postmedieval, Western, male psychology, is the underlying cause of the mistreatment of animals as well as of the exploitation of women and the environment.[3]

Many ecofeminists have argued that the 'postmedieval, Western, male psychology' that Donovan identifies draws stark dualisms between the rational and the natural. That is to say, this psychology separates out rational, scientific mental processes, on the one hand, and natural, physical and emotional feelings, on the other. However, ecofeminists argue that this conceptual framework not only *separates* the rational from the natural, but also *elevates* the former over the latter. In other words, ecofeminists claim that patriarchal culture dictates that reason can and should dominate nature. One needs only to take the briefest of looks at the key thinkers in Western thought to find evidence of this position: Plato equated justice with rule by reason; Aristotle argued that masters can legitimately rule over slaves, and men over women, on the basis of superior reason; Marx argued that man's self-realisation was to be achieved through the mastery of nature and so on.

Importantly, ecofeminists argue that this hierarchical elevation of reason over nature not only provides the rationale for the exploitation and degradation of the environment, but also for the oppression of women. After all, patriarchal norms dictate that women are closer to nature. For example, it has often been claimed that women are more intimately tied to nature because of their role in reproductive processes, because of their mothering instincts and in virtue of their more emotional natures. Because reason is considered to rightfully dominate nature under the patriarchal Western framework, ecofeminists claim that both the environment and women have been oppressed.[4] Donovan's point is that what applies to women and the environment also applies to animals: they too stand on the wrong side of the reason-nature dichotomy, and they too can thus be justifiably dominated under this ideological scheme. As such, these feminist thinkers claim that the liberation of both women and animals depends on overthrowing that scheme.

The second way in which the oppressions of women and animals are claimed to be linked is through the cultural prizing of meat-eating. The way in which meat-eating oppresses animals is fairly obvious, but the link between meat-eating and the oppression of women is less so. However, Carol Adams, in her important book *The Sexual Politics of Meat*, has argued that the practice of meat-eating does have this effect. Adams argues that some human societies evolved so that the consumption of meat became central to their economy and organisation. As such, those that had control over this valuable commodity also came to have elevated power and status. As we know, in most societies, those that controlled access to meat were the hunters, and the hunters were usually men. As such, in economies and societies based around meat, men were given an elevated status over women because of the fact that they hunted and controlled the supply of meat. Importantly, Adams argues that this link between meat-eating and male superiority survives to this day in modern societies. After all, Adams points out that meat-eating is routinely identified with virility, strength and power – but above all, with maleness.[5] Once again then, the liberation of both women and animals depends upon ridding ourselves of this culture of meat-eating.

The third way in which the oppressions of women and animals are said to be linked is through the use of language. Indeed, several theorists have pointed out how animal names are assigned to women as a form of abuse: chick, cow, bitch, dog and so on. Catherine MacKinnon argues that the use of such terms not only degrades women's humanity, but also cements the lower status of animals themselves. Calling a woman an animal assigns her a lower status precisely because animals are recognised as having a lower status. Such name-calling thus serves to oppress both women and animals.[6] Carol Adams has also highlighted how we usually refer to an animal as an 'it', rather than a 'he' or 'she', thus degrading the animal through language by referring to them as things rather than as individuals. Importantly for our concerns, however, Adams points out that our deviations from using 'it' in relation to animals are heavily gendered. For example, when we are referring to an active, dangerous animal that we fear, like a lion, tiger or wolf, we usually refer to 'him', irrespective of their true sex. However, when we are referring to a prey animal that we or other species are hunting, like a deer, fox or even a whale, we usually refer to 'her', irrespective of their true sex. Subtle uses of language like this, so Adams claims, serve to maintain the idea that women and animals are the appropriate subjects of male domination and power.[7] As such, the liberation of women and animals depends upon a thoroughgoing reform of some of the established norms of our language.

The fourth and final way in which both women and animals are claimed to be oppressed is through objectification. Several feminists have pointed out how both women and animals are treated as mere things, as 'pieces of meat' to be possessed and used for the ends of men.[8] Such objectification leads to some of the most cruel and violent acts against women and animals simply for the sexual and gustatory gratification of men:

> ... just as the slaughterhouse treats animals and its workers as inert, unthinking, unfeeling objects, so too in rape are women treated as inert objects, with no attention paid to their feelings or needs. Consequently they feel like pieces of meat.[9]

However, this objectification manifests itself in more subtle ways too. For example, consider the parallels between beauty pageants and animal shows. In both cases, individuals are displayed and paraded as objects to be admired; the sole value of the individuals derives from the excellence of their appearance and the pleasure that their appearance provides for the observer. In both cases, women and animals are reduced to mere things, without a subjectivity, individuality and value of their own. Importantly, this objectification is not merely symbolic or cultural, but instead has been written into the very laws of our societies. After all, both animals and women have been configured as the legal property of men under the law. As such, women and animals are not merely *thought of* as objects, but the law tells us that this is precisely *what* they are.[10] Because of their status as property, the oppression of women and animals is entrenched into societies, subordinating their interests to the interests of their male owners. While women may no longer be the legal property of men in most modern societies, it can be argued that the legacy of this status remains with us today. Once again then, the liberation of both women and animals is argued to depend on an overturning of this objectification.

To conclude this section of the chapter it is necessary to evaluate these claims regarding the linked oppressions of women and animals. Are the oppressions of women and animals interrelated, and are their liberations interdependent as suggested? It is certainly clear that these feminist scholars have unveiled some striking parallels between the oppressions of women and animals. Without doubt, in attempts to maintain their elevated status, white males have certainly employed similar rationales to oppress groups they consider to be inferior, including women and animals. But while knowledge of these similarities is interesting and useful, similarities are not the same thing as inherent connections. Indeed, I want to argue that the oppressions of women and animals are not *necessarily* linked, and thus that their liberations are not *necessarily* interdependent.

For example, it is certainly plausible to claim that an ideological hierarchy elevating reason over nature has been used to oppress both women and animals. However, those oppressions are importantly different. After all, the oppression of women due to their irrationality has been based on a factual error: women are not irrational; women are not of lower intelligence; women are not slaves to their emotions. In an important sense, the irrationality of animals has also been exaggerated. After all, most contemporary biologists and ethologists have provided evidence of remarkably advanced cognitive capacities in a whole range of species of non-human animals. Importantly, however, none of this evidence suggests that these animals are rational *like humans*. It is obviously true that animals are not rational in the way that human beings are rational; there is a qualitative difference in cognitive ability between the two. Quite clearly, however, women are rational *like men*. There is no qualitative difference between the rationality of men and women: men are rational like women are rational. Given this simple fact, we can start to see how the oppressions and liberations of women and animals might be very different. For instance, women might be liberated even when animals remain oppressed. After all, the elevation of reason over nature can be used to maintain the *human* oppression of animals. In other words, it is plausible for women to attain an elevated status because of their reason, while animals remain of lower worth because of their lack of comparable forms of reason. As such, the domination of nature by reason does not then suggest that the oppressions of women and animals are interrelated, and that their liberations are interdependent.

The links between the oppressions of women and animals through meat-eating are also unclear. After all, it seems perfectly possible that meat-eating could be stopped while women remained oppressed; or for women to be liberated while animals continued to be eaten. There is nothing to suggest that societal vegetarianism would necessarily lead to the liberation of women. For one, meat-eating may be associated with maleness, virility and power, but vegetarians can surely also be misogynists. Furthermore, if men gave up control of meat as an economic commodity, they might still retain control over the land, grain, crops and agricultural infrastructure. Quite obviously, economic power does not have to reside in the meat industry. On the other side of the coin, it is also just as possible that women could achieve political and economic power equivalent to men, but continue to eat meat. It is obviously possible that such equality might lead women to actually consume more meat. For in fact, women's liberation might be better facilitated not by demolishing the meat industry, but by women taking over the ownership and management of agribusiness itself. In light of such facts, it is far from clear that the liberations of both animals and women depend on the abolition of meat-eating.

While it is certainly true that some forms of language serve to degrade women and animals, it is once again questionable whether such degradations are necessarily linked. After all, some abusive animal terms apply to men as much as they do to women: weasel, sloth, rat, pig, sheep, donkey and so on are all perfectly clear examples of this. Such terms may cement the lower status of animals, but they do not also serve to degrade women as a whole. Furthermore, some terms which degrade women have nothing to do with animals at all: whore, witch, jezebel, wench and so on. From such facts, we can see that ridding language of all terms which degrade women will not rid language of all terms which degrade animals. And ridding language of all terms which degrade animals will not rid language of all terms which degrade women.

Finally, while it of course must be acknowledged that women and animals are objectified in ways which results in them both suffering terrible harms, the connections between their objectifications are less clear. For example, as acknowledged above, the legal objectification of women through their configuration as property has come to an end in most modern societies. Women are full legal persons under the law in most societies, with all of the rights and responsibilities that flow from that. Women may still be treated or thought of as objects by some men, but they manifestly are *not* objects. Animals, on the other hand, remain very much as objects under the law. The law of modern societies confirms their status as property, thus sanctioning their objectification. It is of course true that societies confer certain obligations upon the owners of animals, limiting what they may permissibly do to their property. Legal objectification does not mean that all and any horrors can be inflicted upon owned animals.[11] But nevertheless, at root the law entrenches a hierarchy between persons and property. Importantly for our purposes, modern societies put women in the former camp, and animals in the latter. As such, the objectification of both must be recognised as different.

To conclude this long section, I refute the claims of some feminist thinkers that the oppressions of women and animals are interrelated, and thus that their liberations are interdependent. While these scholars have done important work in outlining some of the interesting similarities between the ways in which women and animals have been subjugated, they have not proven that their oppressions are intrinsically connected. It is perfectly possible to imagine communities where animals are liberated, but where women are oppressed; and indeed to imagine communities where women are liberated, but animals are oppressed. While there are undoubtedly good reasons to be concerned about and campaign against the oppression of all marginalised groups – women, the working class, animals, blacks, homosexuals, lesbians, transgendered individuals, the disabled and so on – that does not mean that their oppressions are necessarily linked. In practice, liberation for one group does not depend on or entail liberation for all.

The failure of reason

Let us then leave aside the claim that the liberations of women and animals are interdependent. A related but different argument made by care-based theorists is that the liberation of animals (and indeed, women) cannot be achieved by traditional *reason-based* approaches to justice. Indeed, a number of feminist thinkers have been extremely critical of the theories of Peter Singer and Tom Regan, not so much because of the conclusions they draw, but instead because of the methodology that they employ. For these feminist thinkers, Singer, Regan and other mainstream animal ethicists put too much faith in and place too much emphasis on logic and reason to establish our obligations to animals. In fact, some care-based theorists accuse these thinkers of actually fetishising reason in their work, bragging that they rely on well-thought-out arguments and not sentiment to make their case for animal rights. For example, in the preface to *Animal Liberation*, Singer writes as follows:

> This book is an attempt to think through, carefully and consistently, the question of how we ought to treat nonhuman animals. In the process it exposes the prejudices that lie behind our present attitudes and behaviour. In the chapters that describe what these attitudes mean in practical terms – how animals suffer from the tyranny of human

beings – there are passages that will arouse some emotions. These will, I hope, be emotions of anger and outrage, coupled with a determination to do something about the practices described. Nowhere in this book, however, do I appeal to the reader's emotions where they cannot be supported by reason.[12]

For care-based theorists, this valorisation of reason and rejection of sentiment is extremely problematic. In this section of the chapter, I address and analyse five criticisms of reason-based approaches to animal justice that have been made by care-based theorists.

The first claim made against reason-based approaches to animal justice is simply that reason can go awry. To readers of this book, this accusation may have more than a ring of truth about it! After all, the book has analysed a huge range of political thinkers: some put forward very different reasons for withholding justice from animals; others put forward very different reasons for extending justice to animals. Quite simply, not all of these theorists can be right – the vast majority must be entirely wrong! This is perhaps not much of a surprise when we consider that reason is constantly subject to a range of pressures that can make it go awry. As Thomas Kelch writes, 'Reason in sometimes mistaken. Reason can go astray through undue credulity or scepticism, or inappropriate acceptance of authority.'[13]

While there is a certain truth to this criticism, however, we should not make too much of it. Of course reason-based approaches to justice can be wrong. And of course our reasoning can be led astray by false beliefs, poor logic, prejudices, social conditioning and a whole range of other pressures. However, these obvious facts do not mean that we should abandon reason. Rather, they simply imply that we should be humble when formulating our theories and principles. We should be alive to the fact that our reasoning can and will go astray. We should make every effort to stop that from happening. And we must be aware that since it is hopelessly unlikely that we will hit on the absolute truth, we must always be receptive to objections and challenges to our claims in order that they may be improved. But it would be completely self-defeating to abandon reason just because it sometimes goes awry. After all, and quite obviously, we need reason to judge when our reasoning has in fact gone awry!

The second critique that care-based theorists make of reason-based approaches to animal justice is that their claim not to rely on sentimental appeal is in fact false. Care-based theorists argue that all reasonable arguments for extending justice to animals have emotion and feeling lurking behind them, and that it is better to be explicit about the relationship between emotion and morality rather than trying to deny it.[14] Take, for example, the famous so-called 'argument from marginal cases' employed by many mainstream advocates of justice for animals. The argument from marginal cases has popped up at other times throughout this book, even if it has not been explicitly referred to under that title. In effect, the argument runs that we should not withhold justice from animals on the grounds that they lack some characteristic such as rationality, moral agency, autonomy, speech, personhood and so on. For if we were to withhold justice from animals on those grounds, then to be consistent we would also have to withhold justice from human beings who lack those same characteristics, such as young infants and the severely mentally disabled. Since we do want to include such human beings in our considerations of justice, the argument claims that in order to be consistent, we must also include animals with comparable capacities. However, care-based theorists have an interesting response to such an argument. They have pointed out that it only works *if* one assumes that marginal humans *should* be included within justice. What proponents of the argument from marginal cases fail to provide, according to this critique, is a *good reason* for including all humans within justice.[15] Instead, it is claimed that proponents of this argument simply rely on a feeling, a sentiment, an intuition or a gut response, which tells them that young infants and the severely mentally disabled should be included. As a result, what looks like a well-reasoned case for extending justice to animals is actually based on emotional sentiment.[16]

Unfortunately, this critique overlooks the fact that the most important proponents of the argument from marginal cases *have* in fact provided good reasons for extending justice to young infants, the severely mentally disabled and animals. Peter Singer, for example, does not ground his concern for young infants and the severely mentally disabled on an emotional feeling he has towards them, but on the fact that they are sentient beings. It is this capacity for sentience that makes an individual worthy of justice, not any sentimental feeling or gut intuition. Tom Regan provides similar, but different grounds for inclusion. Regan claims that what matters is that an individual is the 'subject-of-a-life', which means that the individual has consciousness, beliefs and desires, perception, memory, the capacity for intentional agency and so on. Once again then, the argument is not that animals should be included just because we feel that humans with similar capacities should be included. Rather, the argument is that animals should be included on the same reasonable grounds that young infants and the severely mentally disabled are included: that they are all 'subjects-of-a-life'.

The third critique of reason-based approaches to animal justice claims that they are out of touch with the views and motivations of those who campaign on behalf of animals. Brian Luke, for example, has put forward this critique eloquently. Luke claims that those campaigners fighting for animal rights are fighting not because they regard what we do to animals as speciesist. That is to say, campaigners for animals are not motivated by the fact that we unreasonably and unfairly treat humans and animals in completely different ways. Instead, Luke points out that they are fighting because they feel that what we do to animals is appalling and disgusting in and of itself.

> My moral condemnation of the acts arises directly from my sympathy for the animals, and is independent of the question of whether humans are protected from such abuse. To the extent that humans are also treated in these ways I object to that, too, but again, out of sympathy, and not considerations of fairness.[17]

Luke's claim is that a sentimental and sympathetic approach to the suffering of animals is what drives the animal protection movement, and that the theories of justice which support that movement should reflect this.

However, there are two important responses to make to this critique. In the first place, it is worth asking just how important this disconnect between the theory behind animal rights and the campaign for animal rights actually is. After all, there are various resources that can be drawn on to extend justice to animals, and it is surely inevitable that campaigners will draw on different resources to motivate them. What unites animal campaigners is that they regard the treatment of animals to be unjust; why they regard it as unjust is an interesting question, but the fact that different campaigners will provide different answers to this question is hardly devastating. Indeed, the same applies to all social movements: for example, some called for the abolition of slavery because they thought it un-Christian, others because it violated the basic rights of men, others because of the suffering it caused and so on. Having an exact fit between the motivations of campaigners and the theories that support social movements seems not only unnecessary, but impossible. Secondly, we must be careful not to exaggerate the differences between animal rights theorists and campaigners. It must be remembered how crucial the works of thinkers like Singer and Regan have been in bolstering the animal movement and bringing in new recruits. Indeed, it is worth remembering that many campaigners actually had their sympathy for animals sparked by the work of these important philosophers. For example, it is common to hear campaigners declare that they first became vegetarian after reading Singer's *Animal Liberation*. Sometimes sympathetic affection for others lies dormant and needs to be prompted in order for it to emerge; the compelling and reasonable arguments of philosophers like Peter Singer and Tom Regan can and do have this effect.

The fourth critique of reason-based approaches to animal justice is mainly concerned with those theories that use rights, like Tom Regan's. Some care-based theorists are sceptical about the use of rights to come up with a theory of justice for animals. They argue that because rights are claims against someone else, they are necessarily antagonistic. For example, Marti Kheel writes, 'The notion of rights can, in fact, be conceived of only within an antagonistic or competitive environment. The concept of competition is inherent in the very definition of rights.'[18] Such thinkers argue that rather than working out what rights animals, and indeed humans, are entitled to, we would be better off working towards a community where such rights are unnecessary. That is to say, we should be working towards a harmonious society in which all can live well without having to make claims against others simply in order to have a decent quality of life.

This argument certainly has some appeal! Who, after all, would be against harmonious communal life? Nevertheless, this critique presents an extremely unfair view of the types of thing that rights are. For one, it unfairly describes rights as *necessarily* antagonistic. For there is nothing inherent in the idea of rights which opposes peaceful community life. Such a life should of course be the goal of societies; a society where no one has any need to claim their rights is a society worth aspiring to. However, inevitably societies regularly fall well short of that aspiration. After all, individuals in all societies through all ages have and continue to act so as to promote their own interests at the expense of others. Because of this simple fact, political theorists argue that we need to have a set of rights to place limits on what individuals can do to one another, and also to delineate what duties individuals have to one another. On this understanding, rights do not *create* antagonisms, but instead attempt to both prevent and remedy the inevitable clashes of interests in society.

The fifth and final critique of reason-based approaches to animal justice argues that they are essentialist: that they reduce animals and their worth to some single, stripped-down core, based around interests. There are in fact four different claims that form this critique, and it is useful to consider each in turn. The first claim is that having some form of consciousness as the single necessary condition for meriting justice renders *relationships* between individuals unimportant in formulating our obligations.[19] Proponents of this position claim that the relationships we have with other individuals are crucial to what we owe them. After all, it is only reasonable to say that we often have different and stronger duties to those who are closer to us, like friends and family. However, this criticism is unfair. After all, it is perfectly possible to use consciousness as a means of deciding who merits justice, but then to let relationships have some say over what obligations we have to particular individuals. Consciousness and relationships are not mutually exclusive concerns. But in any case, and as we discussed in Chapter 5, there are good reasons to limit the role relationships have in formulating the obligations of *political communities*. Relationships may well be crucial for formulating the obligations in our personal lives, but when it comes to delineating the obligations and policies of large societies, some degree of abstraction and impartiality is surely only necessary and right. After all, we have seen throughout history how many communities have excluded from concern groups that they consider to be 'outsiders' – women, blacks, homosexuals, lesbians and so on.

The second claim of the essentialist critique is that by basing animals' worth on their capacity for consciousness, we lose sight of the needs of *particular individual animals*, and instead view them solely as bundles of interests.[20] However, while it is certainly true that animals are not just 'animals', but are instead individuals of particular species, with their own characters and needs, it is worth considering how relevant these facts are for the policies of political communities. After all, societies cannot cater for the personal needs and interests of every single individual, but have to make generalisations about the basic needs and interests of everyone. Just as this is true of human beings, so it is true of animals. One of the basic interests of all sentient animals is to be free from suffering. Formulating policies that serve to protect that interest may generalise away from the more specific needs of particular individuals, but given limited resources and practical considerations, such policies seem perfectly acceptable.

The third related claim of the essentialist critique argues that valuing animals solely in terms of their interests creates hierarchies, where those with the most and strongest interests are elevated above those with fewer and weaker interests.[21] However, we should be extremely wary of the loaded use of the term 'hierarchy' in this context. Reason-based approaches to animal justice do often balance competing interests, and they do make their decisions by coming down in favour of the stronger and more compelling interests; but that is hardly controversial, and nor does it necessarily entail any kind of hierarchy of status. States have to make all kinds of decisions by balancing interests, and this does not necessarily result in a hierarchy of status. For example, when new airports are proposed, political communities have to weigh up a range of competing interests: the interests of those living locally in avoiding noise pollution; the interests of present and future generations in reduced CO_2 emissions; the interests of customers in cheaper and more convenient flights; the interests of businesses in making money; the interests of potential employees; and so on. Coming down in favour of the set of interests that are the strongest and most compelling in such cases only seems sensible. That does not mean that all the other interests under consideration are entirely neglected, or that all groups with weaker interests are somehow of lower status. It is simply that weighing interests and coming down in favour of the strongest offers a plausible, determinate and fair way of making decisions. The same principle applies to decisions over animals. When the interests of animals clash with those of humans, sometimes the interests of humans will be stronger and win out, and sometimes the interests of animals will be stronger and win out. No hierarchy of value needs to result from the process of weighing and balancing interests.

However, the final essentialist critique of reason-based approaches argues that this balancing of competing interests employs the same scientific methodology that has been used to sanction great cruelties to animals.[22] After all, isn't this same logic used to support factory farming and animal experimentation? Both practices have been claimed to be justified on the grounds that they promote the important interests of human beings: the interest in having cheap food, and in having life-saving medicines. But while it is certainly true that arguments along these lines are regularly employed, it is important to question how devastating they are to the very method of balancing interests. After all, thinkers like Peter Singer have pointed out that these types of argument employ *faulty* balancing. Singer's claim, you will recall, is that traditional forms of balancing arbitrarily and unfairly prioritise all the interests of one group (the human species) over all the interests of another group (non-human animals). When we balance interests in a non-speciesist way, Singer claims, radically different proposals follow. As such, many proponents of reasoned-based approaches to animal justice would claim that doing away with balancing of interests on the basis that it has been used unfairly in the past, would be like throwing the baby out with the bathwater.

To summarise then, many care-based feminist theorists are sceptical of the methods employed by such thinkers as Peter Singer and Tom Regan. They argue that these mainstream reason-based approaches to animal justice fail to deliver on their promises because of their faulty methodology. However, I have argued in this section that these critiques are unfair, and that reason-based approaches offer more than these critics think.In any case, an important question for care-based theorists remains: it is all very well criticising reason-based approaches to animal justice, but just what kind of approach should we replace it with? Answering this question is the focus of the next section.

A care-based approach to animal justice

Given the care-based feminist critique of mainstream reason-based theories of animal justice, it is necessary to examine what approach these thinkers propose in its place. Throughout the discussions so far some of the very different ideas in the care-based approach have already been flagged up: the valuing of emotion, sentiment and feeling over reason; the importance of relationships and partiality when delineating obligations; and the importance of context over abstraction when making policy decisions. However, it is necessary to put this care-based approach in a wider and more historical context. After all, understanding the roots of this alternative approach to ethical and political theorising will enable us to better understand its core tenets.

Much of the impetus of the feminist care-based approach derives from the work of Carol Gilligan and her groundbreaking book, *In a Different Voice.*[23] In that book, Gilligan criticised her colleague Lawrence Kohlberg's hierarchical account of moral development. Kohlberg argued that the moral development of human beings comes in six stages. Firstly, children learn to behave in certain ways to avoid punishment and attain reward. Next, children move on to meet the needs of others on the condition that their needs are met in return. Thirdly, adolescents adhere to prevailing norms in order to secure approval. Next, a sense of duty arises out of respect for authority. Fifthly, adults attain a legalistic orientation, which mandates that individuals can do as they like provided that they do not harm others. And finally, adults might adopt a universalist orientation, where they follow self-imposed rules and principles which can be universalised. When Kohlberg assessed the moral development of women and men according to this ladder, women rarely climbed past stage three, whereas men often made it to stages four and five. This result could mean that women are in general less well morally developed than men; or it could suggest that there is a serious problem with Kohlberg's theory! In her book, Gilligan came down resolutely in favour of the latter view, and argued that Kohlberg's theory reflects male bias. Gilligan claimed that Kohlberg's ladder may well represent the moral development of men, but it does not represent the moral development of women. It is inapplicable, according to Gilligan, because women possess an importantly different ethical perspective, a perspective which Kohlberg's methodology takes no account of. Gilligan's own empirical studies on how women confront moral problems revealed that women have a *different* moral voice to men. This voice speaks in the language of *care*; it is a voice that emphasises relationships and responsibilities, rather than abstract principles and rules. Gilligan's claim is not that these differences are set in stone, or that all women think in one way, and all men in another. Rather, her point is that this care-based voice has been neglected by philosophers and educational psychologists, and that this voice should play a prominent role in guiding moral deliberation.[24]

This then is the context in which care-based theorists are now writing. They have taken up the challenge posed by Gilligan, and have been applying this care-based approach to a number of important political and moral concerns. As we have seen, a good number have applied it to the question of how we should treat animals, and have argued that it offers the appropriate means of outlining our obligations to them. One of the earliest and most important applications of this care-based approach to moral and political theorising came from Nel Noddings. In her influential 1984 book, *Caring*, Noddings applied her care-based approach directly to the question of our obligations to animals. Noddings's conclusions in relation to animals reflect the great importance that a care-based approach places in the *relationships* between individuals when formulating our obligations. Noddings maintains that we cannot have an obligation to be a vegetarian, because we cannot be obligated to *all* animals, beyond the bare duty to spare them pain whenever possible.[25] However, we can have obligations to particular types of animal, because we can enter into particular relations with them. For example, Noddings claims that she personally has an ethical responsibility towards cats because she has established a certain type of relation with them: she has welcomed members of that species into her home, enjoyed their company and has cared for them. However, she maintains that because others have had no such relations with cats, they have no such responsibilities. Likewise, Noddings maintains that she has no obligation to rats, beyond refraining from inflicting unnecessary suffering on them, on the basis that she has not established any relations with that type of animal.[26]

Such arguments contain a number of important problems, however, and these problems apply to the care-based concern with relationships and partiality more generally. For example, Noddings's claim that it is necessary to have had a relationship with a particular species in order to have meaningful obligations to members of that species is highly questionable. As was discussed in Chapter 5, there are two very good reasons to be wary of using this kind of partiality when formulating our obligations. In the first place, it can sanction ugly forms of prejudice. After all, if it is necessary to have relations with a particular group in order to have obligations to that group, then it would seem permissible for some communities to discriminate against those they deem 'outsiders' and choose not to mix with. And we have seen countless concrete examples of this, where societies have discriminated against groups on the basis of race, class, gender, religion, sexuality, nationality and so on. Partiality can quickly lead to closed-minded prejudice. As such, many theorists would argue that it is better to ground obligations in the concrete interests of individuals, irrespective of our relations with them. Secondly, when formulating the obligations and policies of *political communities*, it seems only right and proper that those deliberations should be conducted impartially. For it would be wrong and absurd for state officials to assign resources, rights and opportunities according to their own personal relationships and contextual commitments. As was stated in Chapter 5, a degree of partiality may well be proper and permissible in the ethical decisions of individuals, but when formulating the policies of large political communities, it must surely be right to adhere to impartiality as far as possible. In light of these criticisms, I believe that we should be very wary about adopting the emphasis on relationships and partiality that many care-based approaches advocate for. If correct, reason-based approaches, which argue that impartiality requires the extension of justice to animals, may well be more appropriate.

However, a number of care-based theorists have rejected the idea that their theories require narrow-minded partiality. Thinkers like Josephine Donovan have argued that there is no need to limit our care, and thus our obligations, only to those in our most intimate circles. Donovan points out that we do in fact care for far-off strangers. As such, she claims that there is no need to limit our obligations to just those animals we let into our homes as pets. Instead, she argues that we can extend our obligations to include all those animals we care about.[27] However, the obvious objection to this is that peoples' resources of care are not as deep as Donovan might like them to be. For when we look at how much time, effort and resources the vast majority of us give to individuals outside of our immediate group of friends and family – both human and animal – it seems evident that we *do not* in fact care very much about strangers. A care-based approach provides us with no reason to extend our obligations to such individuals, and thus appears an extremely flimsy basis upon which to construct a theory of justice for animals.

Nevertheless, Donovan believes that this objection can be met; and that it can be met by supplementing the care-based approach with a 'political' analysis.[28] Part of that analysis should involve understanding and exposing the different ways politics, institutions, religion, business, culture and the like serve to *limit* our care. For example, Brian Luke argues that caring for animals is in fact the norm for human beings. However, he argues that a huge amount of effort is put into stymieing our natural sympathetic attitudes towards animals. Luke's examples of such efforts include religious stories about man's sanctioned mastery of nature; outright falsehoods from industry about the rationale behind meat production and animal experimentation; and scientific dogmas which state that animals lack consciousness or lack the same type of consciousness that we do.[29] Donovan and Luke argue that a care-based approach must be supplemented by this political analysis in order that our *true* feelings of care for animals can be exposed. As such, they argue that it is perfectly plausible to extend justice to *all* animals using a care-based approach, provided that we take account of our true feelings of care for animals, and not those shaped and moulded by warped religious, political and moral pressures.

But while Donovan and Luke are undoubtedly correct to point out that our attitudes to animals are shaped by political, cultural and other pressures, their analysis of our *true* feelings about animals is highly suspect. Luke, for example, simply stipulates that the natural human response to animals is one of care. The evidence he provides to back up such a bold claim is scant to say the least.[30] And little wonder. For given the diversity of human beings, it is highly doubtful that there is some single natural attitude to animals that all human beings share. Looking at the behaviour of young children, for example, who have not yet been indoctrinated by their family or society, we see that many show great love and affection for animals, while others relish in pulling the tails of cats and tearing the wings off insects. Alternatively, looking at those human communities largely unaffected by the political, religious and corporate pressures that Luke is concerned about, we do find the worship and reverence of animals, but we also find practices such as religious sacrifice that inflict great harms upon animals. Attempting to find some natural human attitude to animals is always going to be forlorn. As such, it is surely better to found our obligations to animals on *good reasons* that have applicability whatever particular feelings individual humans might have about animals. Once again then, it appears wise to hold onto reason-based approaches to the animal issue.

But this is not to say that feeling, sympathy and sentiment have no role to play in morality. After all, it is plainly true, as care-based theorists are keen to point out, that some feeling is required to even begin to consider what obligations we have to others.[31] If humans did not care about anything at all, then there would surely be no means for moral reasoning to get off the ground. However, while it might be appropriate to acknowledge that sympathy is what gets morality going, there are three decisive factors for giving reason the final say when actually delineating our obligations. Firstly, and as we have seen, emotion without reason can sanction blind prejudice. We need reason to help us reflect on our feelings and alert us and inform us when they are inappropriate.[32] Secondly, very often feelings lead in different directions: two equally compassionate individuals can quite plausibly give completely different answers to the same moral question. As such, when a choice needs to be made reason must surely be the final arbiter.[33] Finally, large-scale political communities simply must have general rules in order to function. There must be some set of rules about those forms of behaviour that are permissible and impermissible. And this is true about those forms of behaviour that relate to animals. Importantly, those rules cannot be reconfigured for each and every situation that arises, taking into account our particular emotional responses to it. In order for political communities to function effectively, there must instead be general rules to guide our behaviour. It is my contention that while those rules can be motivated by care, they are best arrived at through the employment of good reasons.

Conclusion

This chapter has examined feminist approaches to the question of animal justice. Because there is no single and definitive form of feminism, and instead a range of different feminist theories, this chapter has focused on that theory which has said most about our obligations to animals: feminist care-based theories. Care-based theories make three important claims concerning the question of justice for animals. First of all, they claim that the oppressions of women and animals are inter-related, and thus that their liberations are interdependent. However, as we have seen, there are good reasons to doubt this claim. It does seem possible to achieve justice for women without also liberating animals; and it does seem possible to achieve justice for animals without also liberating women. The second claim that care-based approaches make is that the liberation of animals cannot be achieved by traditional reason-based approaches to animal justice. They argue that the theories of thinkers like Peter Singer and Tom Regan are fundamentally flawed because, for example, they employ the same methodologies that were used to exclude animals from questions of justice. I have argued, how-ever, that there are good reasons to be sceptical about such critiques of these so-called 'traditional' approaches to animal justice. After all, Singer and Regan have radically modified the philosophical resources they draw upon, providing compelling reasons why traditional theo-ries should be altered, and why they need to take the question of our obligations to animals extremely seriously. Finally, care-based theories argue that the question of animal justice is best approached by a theory that is based on care: which takes the context of a particular situation seriously; which takes into account the importance of relationships to our moral obligations; and which uses feeling and sentiment to delin-eate those obligations. Once again, I have argued that there are good reasons to resist these care-based alternatives when working out what obligations political communities have towards animals. For example, care-based theories have the potential to sanction prejudice and narrow-minded partiality, and possess few resources to help us make difficult choices when our sympathies conflict. Perhaps most importantly, how-ever, care-based theories will always face the difficulty of how to marry their concern for the individual context and particulars of a situation with the need to have general policies for the effective functioning of political communities.

8
Conclusions

This book has introduced and critically examined the most prominent schools in contemporary Western political theory and their implications for our treatment of animals. Now that these different schools are better understood, this concluding chapter aims to look at political theory's relationship with the animal issue *as a whole*. As such, the chapter aims to do three things. Firstly, it says a little about the important contribution that each of the theories makes to the question of how political communities ought to govern their relations with animals. It claims that while each of these theories has its problems, each also adds something important to the debate concerning the question of justice for animals. Secondly, the chapter puts forward an extremely brief defence of utilitarianism and liberalism as the theories which offer the *most* valuable contributions to the debate over our treatment of animals. Finally, in the last section, the chapter makes a few tentative remarks regarding the present state of political theory and its relationship with the animal issue.

The contribution of the prominent schools

The previous chapters spent quite some time outlining the flaws of each of the major schools in Western political theory and the problems with what they have to say about how political communities should treat animals. Quite simply, none of these theories is without its difficulties. For example, utilitarianism can justify sacrificing individual animals and their most important interests, if such sacrifice serves overall welfare. Liberalism can condone cruel practices towards animals by refusing to allow the state to interfere in the private lives of its individual citizens. Communitarianism seeks to promote the politics of the common good, but faces the important difficulty in ascertaining what that good is and who it belongs to. Marxist theory regards the unfolding of history to be solely about the emancipation of human beings, but there is good reason to question whether history is indeed unfolding in such a way. Feminist care-based theories claim to offer strong protections to animals, but given the particular, partial and contextual nature of these theories, might actually condone quite lax standards for those animals people care little about. But in spite of these and other problems, it would be remiss to ignore the important contributions that each of these theories make in relation to the question of animal justice. The remainder of this section defends the view that each theory has something important to offer the debate about what political communities owe to animals.

While the history of political thought is obviously not a distinctive school within political theory, studying that historical context is important when thinking about how contemporary political communities should address their obligations to animals. Chapter 2 claimed that debate about the question of justice for animals has ebbed and flowed throughout the history of Western political thought. There was a rather lively debate about the status of animals in Ancient times. This can be contrasted with the medieval period, in which there existed a consensus over their exclusion from justice. Modern political thought, especially after the eighteenth and nineteenth centuries saw a return of that debate. This historical ebbing and flowing is important to bear in mind when considering how we should govern our political relations with animals. This is because it reminds us that prevalent social, cultural and political norms are not fixed. Rather, those norms change dramatically over time and in response to a range of ideological, religious, environmental and other pressures. It is important then, as political theorists, that we try to abstract from the norms and pressures of time and place, in order that we may consider our obligations to animals more objectively. For example, many of Singer's policy prescriptions in *Animal Liberation* run contrary to the prevailing norms of contemporary societies. Indeed, the mandatory vegetarianism that he proposes clashes drastically with the norms and practices of most affluent modern societies. After all, in such societies, consuming cheap meat and dairy products at every single meal time is simply the norm, and most individuals completely ignore the interests of the animals that provide such food. But it is important to remember that societal norms and practices differ markedly across time and place. History tells us that societies are, have been and can be organised quite differently: the enormous meat consumption of modern affluent societies, for instance, is a novel and unique phenomenon, as the diets of our recent ancestors should remind us. Moreover, the history of political thought tells us that societies do, have and can take the interests of animals far more seriously: Ancient civilizations, for example, did not necessarily consider animals and their interests to be of a necessarily lower and qualitatively different order. Studying history allows us to appreciate such facts. And studying history allows us to judge policies that conflict with the prevalent attitudes of our time and place, such as those of Singer, from a more enlightened perspective. As such, the history of political thought makes an important contribution to the question of justice for animals.

Utilitarianism is the theory that can extend justice to animals most naturally, given both its *welfarism* and its *egalitarianism*. You will recall that utilitarianism is welfarist in that it judges the rightness and wrongness of actions and policies solely in terms of their impact on overall aggregate welfare. Moreover, utilitarianism is egalitarian in that in its pursuit of maximum utility, all individuals with interests count equally. Given that sentient animals can feel pleasure and pain, and thus are very much possessors of interests, utilitarians usually include sentient animals within their theories of justice. I want to argue then that the greatest contribution that utilitarianism makes to the debate about justice for animals is this idea of attributing justice on the basis of *sentience*. Of course, we should not gloss over the fact that attributing justice on the basis of sentience is controversial. Many political thinkers claim that individuals must be more than sentient to be owed justice: they have argued that only 'persons' – rational, moral, autonomous beings – fall within justice. But, and as we have seen, such claims face a grave problem. Quite simply, many humans who are regarded as uncontroversial beneficiaries of justice, such as young infants and the severely mentally disabled, lack the capacities of personhood. In response to this problem, some theorists have offered complex and extremely dubious philosophical arguments – to do with potentiality, threats to just institutions and so on – in order to keep such human non-persons within justice. However, I wish to claim that it is far more straightforward and plausible to include such humans within justice on the simple basis of their sentience. After all, if an individual is sentient then they are a conscious being who can *feel* the effects of others' behaviour. Babies and the severely mentally disabled may not be able to make life goals or ruminate on what morality requires of them, but they can take joy from their lives, and they can suffer in their lives. This capacity makes such individuals quite different from non-sentient entities like plants, rocks or even electricity pylons. Sentient individuals possess lives that can go well or badly *for themselves*. Since sentient individuals lead such lives, it does seem reasonable that political communities should regulate their behaviour towards them for the sake of those individuals themselves. That is to say, it does seem only reasonable that political communities include such individuals within *justice*, whatever it is that justice for them might entail. Furthermore, if political communities should include human non-persons within justice on account of their sentience, then in order to be consistent they must also include sentient animals on account of theirs. This simple connection between sentience and justice is the important and significant contribution that utilitarianism makes to the debate about justice for animals.

The problem with utilitarianism as an account of what we owe to animals, however, is its aggregative nature. By basing each and every decision about a policy on its effects on aggregate utility, utilitarianism neglects the individual. Utilitarianism must even condone the sacrifice of individuals and their important interests when such sacrifice promotes overall welfare. Liberalism avoids this problem by making the individual the prime object of ethical value under its theory. Reminding us to keep the individual at the centre of our theory is the important contribution that liberalism makes with regard to the question of justice for animals. The individual must be kept at the centre of our theory quite simply because in the end, justice has to be about individuals. After all, sentient individuals are the only things that are going to feel and have their lives affected by political principles and policies. Sentience is important to justice, not because pleasure is good in and of itself; but because it is good *for individuals*. This is not to say that justice is entirely individualistic and that communal attachments have no importance. It is simply to remember that such communal attachments are only of value in so far as they benefit individuals. At bottom, justice must never neglect individuals, because justice must in the end always be about individuals. This prizing of the individual is the important contribution that liberalism makes to the question of how political communities should govern their relations with animals.

Communitarianism takes issue with this prizing of the individual and aims to put communal norms at the centre of its view of justice. However, communitarianism is overly optimistic about the cohesion of values and norms within communities. As such, while the community must obviously be of crucial significance to any theory of justice, it must be remembered that the community is of importance only because it is of value to individuals. In spite of this critique of communitarianism, the theory nevertheless makes an important contribution to the debate about the question of what political communities owe to animals. This is because communitarianism reminds us that the individual can only flourish given an appropriate and supportive societal context. For example, communitarians often point out that the liberal ideal of all citizens being able to frame, revise and pursue their own conceptions of the good is only possible in a particular type of society. That is to say, it is only possible within a society that provides the education, resources and opportunities for such autonomous living. This insight has two important implications for the question of justice for animals. First of all, it reminds us that many animals are social creatures too, whose welfare often depends upon being members of a wider and flourishing social group. Secondly, it further reminds us that the protection of individual sentient animals will only be possible in a society that supports such animal protection. In other words, a large part of achieving justice for animals requires a shift in societal attitudes about the worth of animals and their interests. As we can see then, communitarianism provides the valuable insight that the flourishing of individual animals depends upon certain forms of communal support.

The idea that achieving justice for animals requires certain fundamental shifts in society is also shared by Marxism. In fact, the great contribution that Marxism makes to the debate about the proper treatment of animals derives from the way in which it measures success, and the insights it provides on how to achieve that success. What I mean by this is that Marxism recognises that the simple enactment of legislation will never achieve justice for individuals, no matter how noble, virtuous or just that legislation is. For Marxists, no real justice can ever be achieved while we live under oppressive capitalist relations. We can change the laws of our states all we like, but so long as we live in conditions of scarcity, it is argued, where some have control over resources and others do not, justice can never be achieved. As such, Marxists claim that fundamental shifts in economic and societal relations are required in order to liberate individuals. Now, we do not have to go as far as Marxists and demand the overthrow of capitalism to acknowledge the wisdom of the point being made here. For the simple point is that *legal* change does not necessarily equate with *effective* change. For example, we can institute stronger animal welfare laws, and we can even grant animals legal rights akin to those afforded to human beings, but none of this guarantees achieving justice for animals. For in order for animals to actually *enjoy* their rights, fundamental shifts in the organisation, norms and institutions of society are required. As mentioned above, I do not believe that those shifts need necessarily entail the overthrow of capitalism. However, they certainly necessitate a move away from capitalism as currently practised, where the interests of animals count for almost nothing. If animals are to enjoy their rights, they need to be regarded as sentient individuals who merit justice for their own sake, and not as mere commodities to be raised and killed as efficiently as possible. Legal action can obviously help to effect great changes for animals, but not if it is unaccompanied by fundamental shifts in societal norms and practices alongside. The great contribution that Marxism makes to the debate about justice for animals is the notion that legislative change does not guarantee the achievement of justice for individual animals.

The final theory addressed in the book, care-based feminism, also has important things to say about how political communities should govern their relations with animals. While I was quite critical in Chapter 7 about using emotion and sentiment as a means of determining the principles and policies of political communities, that does not mean that such feelings should be entirely neglected when addressing our obligations to animals. After all, sympathy for the suffering of others often alerts us to injustice and motivates us to effect change. As such, the use and manipulation of sentimental feeling must be a key part of the strategy of activists who campaign on behalf of animals, and indeed who seek to effect the changes of societal norms and institutions referred to above. Even 'rationalist' thinkers like Peter Singer would have to admit this. For while Singer denies in *Animal Liberation* that his case rests on an appeal to sentiment, the pictures of animal suffering that he includes within the book, and the descriptive accounts he provides of the cruelties they face in laboratories and farms, cannot fail to move. Sentiment and emotion are key elements of any moral enterprise, and must form part of the means of achieving justice for animals, as they must for the achievement of justice more widely. Nevertheless, and to reiterate, that does not mean that emotion should tell us what the *content* of justice is. For as argued in the previous chapter, emotion can be indeterminate, prejudicial and biased. For these reasons we should be extremely wary of using it as a basis for making general decisions about how political communities ought to treat animals.

The wisdom of utilitarianism and liberalism

It will probably be clear from the above that the political theories I am most sympathetic to in general, and when formulating our obligations to animals in particular, are utilitarianism and liberalism. While there is obviously not the space here to outline and defend a full theory of justice for animals which draws on these two theories, I do think that it will be useful to say something about what I believe these theories offer, and how I believe that the two schools can be married. Hints of this position can be found in the preceding chapters, most notably in Chapters 3 and 4, but I hope it will be beneficial to set out these claims here more explicitly.

As noted above, the great wisdom of utilitarianism is the importance it gives to sentience when deciding who merits justice. In devising actions, policies, principles and institutions aimed at maximising utility, all individuals with interests count. I have argued that this way of attributing justice is plausible not so much because sentience itself is important, but because of what sentience represents. An individual with sentience is an individual with a life that can go well or badly for themself; an individual with sentience necessarily has a well-being that the actions and policies of others can set back or promote. As such, sentient individuals are quite unlike inanimate entities such as rocks, plants and electricity pylons in that sentient individuals can feel for themselves the effects of others' actions. Sentient individuals are thus extremely strong candidates for the type of being that political communities have obligations towards. However, while the utilitarian focus on welfare as a means of attributing justice is plausible and admirable, its focus on *aggregate* welfare as the means of defining justice is not. As emphasised repeatedly, by focusing on aggregate welfare, utilitarianism too readily loses sight of the individual. I wish to argue that utilitarians grant too much value to welfare per se, forgetting that welfare must ultimately be valuable in so far as it matters to *individuals*. Sentient individuals merit justice not because they can create more welfare in the universe, but because they can feel the effects of particular actions and policies. It is the individual subject of welfare that is of ultimate ethical importance, not the total amount of welfare in the universe. Liberal theory reminds us of the value of individuals and as such acts as a useful check on utilitarian theory. However, as we saw in Chapter 4, traditionally, liberalism does not value individuals per se, but individual *persons*. Since most animals are not individuals who can act morally, nor are they individuals who can frame, revise and pursue their own conceptions of the good, many liberals have excluded, and continue to exclude, animals from justice. In summary, traditional utilitarian theory offers the advantage of welfarism, but the disadvantage of aggregate welfarism; while liberal theory offers the advantage of valuing individuals, but the disadvantage of only valuing individual persons. Can there be a theory that encompasses the great benefits of these two theories, while also excluding their costs? Is an individualistic welfarism possible?

I want to argue that it is. It is perfectly possible for political communities to acknowledge the wisdom of utilitarian theory and recognise that their obligations extend to all sentient beings, but also to acknowledge the wisdom of liberalism and recognise that those obligations are owed to individuals themselves, and may not always be aimed at maximising overall utility. One way for political communities to do this was described in Chapter 3, and that would be for political communities to assign 'interest-based rights' to animals. Such rights would limit what can be done to sentient individuals, even in the name of overall utility. The idea is that rights act as a kind of protective shield for individuals, demarcating the kinds of action that no individual should be subjected to, no matter what benefits to society might accrue. This puts a stop to the kind of utilitarian policies that might sacrifice individuals and their interests for the greater good. As such, just as we do not experiment on or intensively farm *human beings*, no matter what overall benefits might be afforded, so we could decide not to experiment on or intensively farm *animals*, irrespective of the benefits of animal experimentation and intensive farming. Political communities can limit the range of permissible actions in relation to animals, by assigning them rights that are grounded in their individual interests.

But interest-based rights do not only place limits on the pursuit of aggregate utility, for they can also place limits on the exercise of *freedom* within liberal political communities. Just as John Stuart Mill claimed that the liberty of citizens can legitimately be constrained in order to prevent harm to other human beings, so it can be legitimately constrained in order to prevent harm to other animals.[1] Cruelty to animals is not a private, domestic, religious or moral issue that should be protected from scrutiny. Just as spousal abuse and child abuse are legitimate concerns of the wider political community, so too is the abuse of animals. Moreover, just as institutionalised forms of cruelty to humans – as in trafficking and slavery – are not just left to the whims of the marketplace, neither too should institutionalised forms of cruelty to animals, such as factory farming and animal experimentation. Rights can be assigned to animals to protect their basic interests from the exercise of liberty by citizens and property owners, as well as from the communal pursuit of overall utility.

Of course, many political thinkers have difficulty with the idea that political communities should assign rights to animals on the basis of their interests. Such sceptics would argue that such a policy would lead to far too many rights claims, and to far too many clashes of rights.[2] After all, animals – and humans – have interests in all sorts of things, and awarding rights to all of those interests would surely lead to ever increasing and ever implausible sets of rights claims. For example, my cat clearly has an interest in sleeping on my lap, but surely it would be absurd to argue that he should thus have a corresponding *right* to sleep on my lap that can be enforced by the state. Basing rights on interests, such objectors would claim, leads to an unwelcome and implausible inflation of rights.

However, this objection wrongly assumes that an interest-based theory of rights must turn each and every interest into a right. Interest-based theories of rights claim that all rights derive from interests, but they do not claim that all interests should become rights. Any plausible interest-based theory of rights must *discriminate* between interests in order to show why some interests afford the protection of rights, and others do not. There is not the space here to offer a fully worked out theory of how that might be done, but it is worth pointing out that such discrimination is perfectly possible. For example, the strength of the particular interest at stake will surely be relevant to whether a right is ascribed. Indeed, it seems fairly obvious that the interest of my cat in sleeping on my lap is far too trivial to warrant the award of a corresponding legal right to him. However, his interest in not being beaten for my amusement is clearly much stronger. As such, awarding him, and other pet cats, a legal right not to be beaten is certainly far from fanciful. Other cases, like the interests of animals not to be killed to be eaten, or in not being subjected to painful medical experiments, or in not having their habitats destroyed by human development projects, are certainly more difficult, and each will require careful consideration of the strength of the relevant interests at stake. But there is no reason why an interest-based theory of rights cannot provide determinate conclusions regarding these or any other situations.

I want to claim that this interest-based rights approach offers the appropriate fusion of liberal and utilitarian theories, which in turn offers the most appropriate way of delineating political communities' obligations to animals. However, it is extremely important to note that this approach by no means guarantees the *achievement* of justice for animals. Assigning legal rights to animals on the basis of their interests will be impossible and ineffective unless there are fundamental shifts in societal attitudes, norms, practices and institutions. Here the wisdom of communitarianism, Marxism and feminist care-based theories must be admitted. While hope of a shift to a *consensus* on the value of animals and their interests is both futile and unnecessary, it is certainly true that achieving justice for animals requires contemporary political communities to take the animal issue much more seriously. Communities do need to recognise that the interests of animals count for *something*, and do need to realise that this simple fact will have important implications for many of their practices. Realising such shifts in societal attitudes, norms, practices and institutions will of course be facilitated by the good arguments of political theorists, but it of course needs to be supplemented by the sentimental and emotional engagement of animal activists.

Contemporary political theory and animals

But if the good arguments of political theorists are needed in order for political communities to take the animal issue more seriously, we need to examine to what extent they are being offered. Is the contribution that I have offered in this chapter part of a lively debate within political theory about our obligations to animals, or is it an eccentric voice within a discipline concentrating on very different issues?

Putting things in a historical context, it certainly has to be acknowledged that we are currently living in the most fertile period that there has ever been regarding discussions about political communities' obligations to animals. As I have stated elsewhere in the book, the publication of *Animal Liberation* in the 1970s resulted in a series of responses and critiques from political thinkers working within very different philosophical frameworks. As such, there now exist a number of quite distinctive theoretical approaches to the question of justice for animals, as the chapters of this book provide evidence of. Furthermore, publications on what is owed to animals are commonplace in mainstream journals and publishing houses, and political thinkers have been and continue to be involved in helping frame animal welfare legislation across many states.

But in spite of this activity, it does have to be recognised that the discussion of what we owe to animals remains peripheral within contemporary political theory. The number of political theorists who have made contributions to the debate is still relatively small, and most new contributions to debates about social justice per se make no mention of obligations to animals whatsoever. While the issue of whether justice is owed to animals may no longer be considered as entirely absurd, it is nevertheless still considered as something of an oddity by many political theorists. This neglect of the animal issue by most political theorists, however, is hardly surprising. After all, political theorists work within and are informed by the societal norms and values of their day, and it is clear that the norms and values of most modern societies have little regard for the interests of animals. It is common to hear that concern for animals is somehow childish, emotional or trivial. Our ethical concern and energies, we are often told, are better directed towards the real harms suffered by human beings, because it is humans and humans alone that are of ultimate ethical importance. Given that political theorists want to be taken seriously, both as academics and as commentators on the pressing political issues of the day, it is little wonder that so few have dared to turn their attention to issues as allegedly trivial as the protection of animals.

However, this neglect is to the detriment of political theory. It is of vital importance that political thinkers shed themselves of the ideological straightjackets of their societies, and instead consider how political communities should be organised from a more enlightened perspective. Abstracting from prevalent social norms is the only way that theorists can be sure to avoid reinforcing prejudices and unjustified biases. Importantly, the exalted status given to human beings in contemporary Western politics and morality is just such an unjustified bias. After all, we have known for centuries that human beings do not inhabit the centre of the universe. We have also known at least since Darwin's time that human beings are not qualitatively different from other species, but share common ancestry with all forms of life. Finally, it is also perfectly obvious that sentient animals are part of our societies and feel the effects of our policies and actions. Given these simple facts, it seems only right that political communities regulate their behaviour towards sentient animals. Indeed, it seems only right that political communities show sentient animals *justice*.

Now, if all of this is correct, the animal issue is far from a trivial concern, and should move from the periphery to the mainstream of political theory. For if we owe something to animals on the basis that they are sentient and can be affected by our actions, then certain of the ways in which we treat them must come under urgent review. If animals and their interests count for anything at all, then the massive and routine harms inflicted on animals in food production, medical experimentation and habitat destruction must count amongst the most pressing concerns facing political communities today.

Notes

1 Introduction: Animals and Political Theory

1. For reasons of style and clarity, throughout the book I generally follow the norm of using the term 'animals' when I am actually referring to 'non-human animals'.
2. For further details, see http://www.defra.gov.uk.
3. L. Glendinning, 'Spanish Parliament Approves "Human Rights" for Apes', *Guardian*, 26 June 2008.
4. J. Watts, 'China Plans First Animal Welfare Law', *Guardian*, 16 June 2009.
5. M. Bittman, 'Rethinking the Meat-Guzzler', *New York Times*, 27 January 2008.

2 Animals in the History of Political Thought

1. Ovid, *Metamorphoses*, trans. D. Raeburn (London: Penguin, 2004), Book 15.
2. R. Sorabji (1993) *Animal Minds and Human Morals: The Origins of the Western Debate* (New York: Cornell University Press), p. 208.
3. R. Preece (2008) *Sins of the Flesh: A History of Ethical Vegetarian Thought* (Vancouver: UBC Press), Ch. 3.
4. B. Russell (1994) *History of Western Philosophy: and its Connection with Political and Social Circumstances from the Earliest Times to the Present Day*, 2nd edn (London: Routledge), p. 50.
5. Plato, *The Republic*, trans. D. Lee, 2nd edn (London: Penguin, 2003).
6. Ibid., Book II, 372 a–d.
7. D.A. Dombrowski (1984) *The Philosophy of Vegetarianism* (Amherst, MA: University of Massachusetts Press), Ch. 4.
8. Plato, *The Republic*, Book II, 373 c.
9. G.R. Carone (2001) 'The Classical Greek Tradition' in D. Jamieson (ed.) *A Companion to Environmental* Philosophy (Oxford: Blackwell), p. 71.
10. Aristotle, *The Politics*, trans. T. Sinclair, revised edn (London: Penguin, 1992), Book I, 1253 a7.
11. Ibid.
12. Aristotle, *The Nicomachean Ethics*, trans. D. Ross (Oxford: Oxford University Press, 1991), Book VIII, 1161a–b.
13. Sorabji, *Animal Minds and Human Morals,* p. 124.
14. Ibid., pp. 174–5.
15. G. Steiner (2005) *Anthropocentrism and its Discontents: The Moral Status of Animals in the History of Western Philosophy* (Pittsburgh, PA: University of Pittsburgh Press), p. 96.

16. Porphyry, *On Abstinence from Animal Food,* trans. T. Taylor (London: Centaur Press, 1965), Book III, 8.
17. Steiner, *Anthropocentrism and its Discontents,* p. 97.
18. For the contemporary debate, see G. Francione (2008) *Animals as Persons: Essays on the Abolition of Animal Exploitation* (New York: Columbia University Press).
19. Porphyry, *On Abstinence from Animal Food,* Book III, 19.
20. Augustine, *The City of God Against the Pagans,* trans. R. Dyson (Cambridge: Cambridge University Press, 1998), Book I, 20.
21. T. Aquinas, *Summa Theologica,* trans. Fathers of the English Dominican Province (London: Burns Oates & Washbourne Ltd, 1929), Part 2, 25.3.
22. T. Aquinas, *Summa Contra Gentiles,* trans. The English Dominican Fathers (London: Burns Oates & Washbourne Ltd, 1928), Book III, 113.
23. J. Passmore (1975) 'The Treatment of Animals', *Journal of the History of Ideas,* 36, 195–218, especially pp. 197–202.
24. R. Preece and L. Chamberlain (1993) *Animal Welfare and Human Values* (Waterloo, Ontario: Wilfrid Laurier University Press), p. 5.
25. See L. White (1967) 'The Historical Roots of Our Ecological Crisis', *Science,* 155, 1203–7.
26. See A. Linzey (1994) *Animal Theology* (London: SCM Press).
27. Augustine, *The City of God Against the Pagans,* Book I, 20.
28. Aquinas, *Summa Theologica,* Part 2, 25.3.
29. Steiner, *Anthropocentrism and its Discontents,* p. 124.
30. Ibid., p. 125.
31. N. Machiavelli (1532) *The Prince,* ed. Q. Skinner and R. Price (Cambridge: Cambridge University Press, 1988).
32. There is some debate concerning whether Descartes denied feeling to animals. For a persuasive defence of the 'conventional' position that he did deny them such a capacity, see Steiner, *Anthropocentrism and its Discontents,* Ch. 6.
33. T. Hobbes (1651) *Leviathan,* ed. C. Macpherson (London: Penguin, 1968).
34. T. Hobbes (1651) *De Cive* in B. Gent (ed.) *Man and Citizen* (Indiana, IN: Hackett Publishing, 1991), Book 8, 10.
35. J. Locke (1690) *Two Treatises on Government,* ed. P. Laslett, student edn (Cambridge: Cambridge University Press, 1988).
36. J-J. Rousseau (1762) *The Social Contract,* trans. M. Cranston (London: Penguin, 1968).
37. J-J. Rousseau (1755) *A Discourse on Inequality,* trans. M. Cranston (London: Penguin, 1984), Part 1.
38. I. Kant (1963) *Lectures on Ethics,* trans. L Infield (New York: Harper and Row), p. 239.
39. For an example of this argument, see J. Griffin (2008) *On Human Rights* (Oxford: Oxford University Press), Ch. 4.
40. See A. Garrett (2007) 'Frances Hutcheson and the Origin of Animal Rights', *Journal of the History of Philosophy,* 45, 243–65.
41. H. Primatt (1992) *The Duty of Mercy and the Sin of Cruelty to Brute Animals,* ed. R. Ryder and J. Baker (Fontwell, Sussex: Centaur Press).
42. J. Bentham (1780) *An Introduction to the Principles of Morals and Legislation* (Oxford: Clarendon Press, 1823), Ch. 17.

3 Utilitarianism and Animals

1. R. Harrison (1964) *Animal Machines: The New Factory Farming Industry* (London: V. Stuart).
2. P. Singer (1995) *Animal Liberation*, 2nd edn (London: Pimlico).
3. P. Singer (1993) *Practical Ethics*, 2nd edn. (Cambridge: Cambridge University Press), p. 63.
4. Ibid., pp. 64–5.
5. I borrow the terminology of 'therapeutic and non-therapeutic experiments' from R. Garner (1993) *Animals, Politics and Morality* (Manchester: Manchester University Press), p. 121.
6. Singer, *Practical Ethics*, p. 67.
7. Ibid.
8. H. McCloskey (1965) 'Rights', *The Philosophical Quarterly*, 15, 115–27, especially, p. 126.
9. R. Frey (1980) *Interests and Rights* (Oxford: Clarendon Press), p. 82.
10. L. Petrinovich (1999) *Darwinian Dominion: Animal Welfare and Human Interests* (London: MIT Press), p. 217.
11. Ibid., pp. 3–4.
12. This point can be found in P. Devine (1978) 'The Moral Basis of Vegetarianism', *Philosophy*, 53, 481–505, especially p. 485; and T. Regan (1980) 'Utilitarianism, Vegetarianism and Animal Rights', *Philosophy and Public Affairs*, 9, 305–37, especially, pp. 309–11.
13. This statistic is taken from G. Matheny (2005) 'Utilitarianism and Animals' in P. Singer (ed.) *In Defense of Animals: The Second Wave* (Oxford: Blackwell), p. 13.
14. P. Singer (1987) 'Animal Liberation or Animal Rights?', *The Monist*, 70, 3–14, especially, p. 9.
15. Singer, *Practical Ethics*, p. 134.
16. S. Davis (2003) 'The Least Harm Principle May Require that Humans Consume a Diet Containing Large Herbivores, Not a Vegan Diet', *Journal of Agricultural and Environmental Ethics*, 16, 387–97; G. Schedler (2005) 'Does Ethical Meat Eating Maximize Utility?', *Social Theory and Practice*, 31, 499–511.
17. G. Matheny (2003) 'Least Harm: A Defense of Vegetarianism From Steven Davis's Omnivorous Proposal', *Journal of Agricultural and Environmental Ethics*, 16, 505–11, especially, p. 506.
18. M. Nussbaum (2006) *Frontiers of Justice: Disability, Nationality, Species Membership* (London: The Belknap Press of Harvard University Press), p. 345.
19. This example is borrowed from S. Cataldi (2002) 'Animals and the Concept of Dignity: Critical Reflections on a Circus Performance', *Ethics and the Environment*, 7, 104–26.
20. Nussbaum, *Frontiers of Justice*, p. 347.
21. See, for example, P. Taylor (1986) *Respect for Nature: A Theory of Environmental Ethics* (Princeton, NJ: Princeton University Press).
22. Nussbaum, *Frontiers of Justice*, p. 361.
23. Ibid., p. 366.
24. P. Singer, 'A Response to Martha Nussbaum', http://www.utilitarian.net (accessed 18 June 2010).

25. T. Regan (2004) *The Case for Animal Rights*, 2nd edn (Berkeley, CA: Berkeley University Press), p. 246.
26. Ibid., p. 249.
27. The two most famous proponents of this view are M. Fox (1978) 'Animal Liberation: A Critique', *Ethics*, 88, 106–18; and C. Cohen (1986) 'The Case for the Use of Animals in Biomedical Research', *New England Journal of Medicine*, 315, 865–70.
28. Ibid.
29. N. Nobis (2004) 'Carl Cohen's "Kind" Arguments For Animal Rights and Against Human Rights', *Journal of Applied Philosophy*, 21, 43–59.
30. Regan, *The Case for Animal Rights*, p. 243.
31. Ibid., p. 78.
32. See Regan's discussion of this in ibid., pp. 243–48.
33. J. Feinberg (1974) 'The Rights of Animals and Unborn Generations' in W. Blackstone (ed.) *Philosophy and Environmental Crisis* (Athens, GA: University of Georgia Press).
34. For one example of an interest-based theory of rights applied to animals, see A. Cochrane (2007) 'Animal Rights and Animal Experiments: An Interest-based Approach', *Res Publica*, 13, 293–318.
35. See J. Waldron (2005) 'Torture and Positive Law: Jurisprudence for the White House', *Columbia Law Review*, 105, 1681–750.

4   Liberalism and Animals

1. J. Rawls (1999) *A Theory of Justice*, revised edn (Oxford: Oxford University Press), p. 53.
2. Ibid., p. 441.
3. Ibid., p. 442.
4. Ibid., p. 96.
5. Ibid., p. 442.
6. Ibid.
7. Ibid., p. 448.
8. Ibid.
9. Ibid., p. 446.
10. Ibid., p. 443.
11. P. Carruthers (1992) *The Animals Issue: Moral Theory in Practice* (Cambridge: Cambridge University Press), pp. 114–15. Robert Elliot offers the same interpretation of Rawls's argument in R. Elliot (1984) 'Rawlsian Justice and Non-human Animals', *Journal of Applied Philosophy*, 1, 95–106, especially, p. 96.
12. Carruthers, *The Animals Issue*, p. 117.
13. On this issue of the moral personality of animals, see, for example, M. Bekoff and J. Pierce (2009) *Wild Justice: The Moral Lives of Animals* (Chicago, IL: University of Chicago Press).
14. Rawls, *A Theory of Justice*, p. 38.
15. R. Garner (2003) 'Animals, Politics and Justice: Rawlsian Liberalism and the Plight of Non-humans', *Environmental Politics*, 12, 3–22, especially, pp. 16–17.
16. R. Abbey (2007) 'Rawlsian Resources for Animal Ethics', *Ethics and the Environment*, 12, 1–22, especially, pp. 5–6.

17. M. Pritchard and W. Robison (1981) 'Justice and the Treatment of Animals: A Critique of Rawls', *Environmental Ethics*, 3, 55–61, especially, p. 56.
18. For versions of this argument, see D. VanDeVeer (1979) 'Of Beasts, Persons, and the Original Position', *The Monist*, 62, 368–77; Elliot, 'Rawlsian Justice and Non-human Animals'; B. Singer (1988) 'An Extension of Rawls' Theory of Justice to Environmental Ethics', *Environmental Ethics*, 10, 217–31; M. Rowlands (1997) 'Contractarianism and Animal Rights', *Journal of Applied Philosophy*, 14, 235–47; T. Regan (2004) *The Case for Animal Rights*, 2nd edn (Berkeley, CA: University of California Press), pp. 164–71.
19. This claim is made by VanDeVeer, 'Of Beasts, Persons, and the Original Position', p. 372; by Rowlands, 'Contractarianism and Animal Rights', p. 245; and by Regan, *The Case for Animal Rights*, p. 171.
20. B. Barry (1989) *Theories of Justice* (Hemel Hempstead: Harvester-Wheatsheaf), p. 212.
21. Elliot, 'Rawlsian Justice and Non-human Animals', p. 103; Singer, 'An Extension of Rawls' Theory of Justice to Environmental Ethics', p. 223.
22. Rawls, *A Theory of Justice*, pp. 15–18.
23. Garner, 'Animals, Politics and Justice: Rawlsian Liberalism and the Plight of Non-humans', p. 10.
24. Ibid.
25. For such claims, see P. Cavalieri and P. Singer (1993) (eds) *The Great Ape Project: Equality Beyond Humanity* (London: Fourth Estate).
26. Regan, *The Case for Animal Rights*, pp. 84–5.
27. Ibid., p. 249.
28. S. Wise (2002) *Unlocking the Cage: Science and the Case for Animal Rights* (Oxford: Perseus Press), p. 32; E. Pluhar (1995) *The Moral Significance of Human and Nonhuman Animals* (Durham, NJ: Duke University Press), p. 249.
29. G. Francione (2004) 'Animals: Property or Persons?' in C. Sunstein and M. Nussbaum (eds) *Animal Rights: Current Debates and New Directions* (Oxford: Oxford University Press), pp. 124–5.
30. G. Francione (1996) *Rain Without Thunder: The Ideology of the Animal Rights Movement* (Philadelphia, PA: Temple University Press), p. 18.
31. Ibid., pp. 154–5.
32. For more on these points, see A. Cochrane (2009) 'Do Animals Have an Interest in Liberty?', *Political Studies*, 57, 660–79.
33. R. Garner (2005) *The Political Theory of Animal Rights* (Manchester: Manchester University Press), p. 44.
34. On these points, see A. Cochrane (2009) 'Ownership and Justice for Animals', *Utilitas*, 21, 424–42.
35. Ibid.; Garner, *The Political Theory of Animal Rights*, p. 46.
36. R. Dworkin (1973) 'The Original Position', *University of Chicago Law Review*, 40, 500–33, quote, p. 532.
37. J.S. Mill (1859) *On Liberty* in J. Gray (ed.) *John Stuart Mill: On Liberty and Other Essays* (Oxford: Oxford University Press, 1998), p. 14.

5   Communitarianism and Animals

1. The important works of these thinkers include M. Walzer (1983) *Spheres of Justice: A Defense of Pluralism and Equality* (New York: Basic Books); M. Sandel

(1982) *Liberalism and the Limits of Justice* (Cambridge: Cambridge University Press); A. MacIntyre (1981) *After Virtue: A Study in Moral Theory* (London: Duckworth); C. Taylor (1985) *Philosophy and the Human Sciences* (Cambridge: Cambridge University Press).

2. 'Chef Faces £86,000 Chicken Bill', *BBC News Online*, 8 June 2008, available at http://news.bbc.co.uk/1/hi/business/7442601.stm (accessed 18 June 2010).

3. George Orwell quoted in J. Serpell and E. Paul (1994) 'Pets and the Development of Positive Attitudes to Animals' in A. Manning and J. Serpell (eds) *Animals and Human Society: Changing Perspectives* (London: Routledge), pp. 127–8.

4. Walzer, *Spheres of Justice*, p. 313.

5. S. Moller Okin, J. Cohen, M. Howard and M. Nussbaum (1999) *Is Multiculturalism Bad for Women?* (Princeton, NJ: Princeton University Press).

6. B. Parekh (2000) *Rethinking Multiculturalism: Cultural Diversity and Political Theory* (Basingstoke, Hampshire: Palgrave), p. 136.

7. E. Reinders (2000) 'Animals, Attitude Toward: Buddhist Perspectives' in W. Johnson (ed.) *Encyclopedia of Monasticism Vol. 1* (Chicago, IL: Fitzroy Dearborn Publishers), pp. 30–1.

8. Serpell and Paul, 'Pets and the Development of Positive Attitudes to Animals', p. 127.

9. IPSOS-MORI (2009) 'Views on Animal Experimentation', http://www.ipsos-mori.com/Assets/Docs/Polls/views-on-animal-experimentation-2009.pdf (accessed 18 June 2010).

10. DEFRA (2007) 'Survey of Public Attitudes and Behaviours Toward the Environment: 2007', http://www.defra.gov.uk/evidence/statistics/environment/pubatt/index.htm (accessed 18 June 2010).

11. Statistics from the Gallup poll for Realeat show vegetarians made up 2.1 per cent of the population in 1984, rising to a peak of 5.4 per cent in 1994, and falling to 4 per cent in 2001. These statistics are available from the Vegetarian Society's website, http://www.vegsoc.org.

12. M. Midgley (1983) *Animals and Why They Matter* (Athens, GA: University of Georgia Press), p. 102.

13. Ibid., p. 26.

14. Ibid., p. 23.

15. Ibid.

16. Ibid., Ch. 10.

17. Ibid., p. 98.

18. E. Aaltola (2005) 'Animal Ethics and Interest Conflicts', *Ethics and the Environment*, 10, 21–48, especially, p. 31.

19. Ibid.

20. On this distinction between first- and second-order impartiality, see B. Barry (1995) *Justice as Impartiality* (Oxford: Oxford University Press), p. 194.

21. J. Baird Callicott (1989) 'Animal Liberation and Environmental Ethics: Back Together Again' in J. Baird Callicott, *In Defense of The Land Ethic: Essays in Environmental Philosophy* (Albany, NY: State University of New York Press), p. 55.

22. Ibid., p. 58.
23. For the classic argument in favour of helping starving strangers, see P. Singer (1972) 'Famine, Affluence, and Morality', *Philosophy and Public Affairs*, 1, 229–43.
24. J. Baird Callicott (1998) ' "Back Together Again" Again', *Environmental Values*, 7, 461–75, especially, p. 470.
25. Y.S. Lo (2001) 'The Land Ethic and Callicott's Ethical System (1980–2001): An Overview and Critique', *Inquiry*, 44, 331–58, especially, pp. 349–53.
26. J. Hadley (2007) 'Critique of Callicott's Biosocial Moral Theory', *Ethics and the Environment*, 12, 67–78, especially p. 75.
27. For current government policy and its response to FAWC's recommendations, see http://www.defra.gov.uk.
28. C. Kukathas (1997) 'Cultural Toleration' in I. Shapiro and W. Kymlicka (eds) *NOMOS XXIX: Ethnicity and Group Rights* (New York: New York University Press), p. 87.
29. J. Williams, 'US Judge Blocks Tribal Whaling', *BBC News Online*, 4 May 2002, available at http://news.bbc.co.uk/1/hi/world/americas/1967677.stm (accessed 18 June 2010).
30. The same argument is made in B. Barry (2001) *Culture and Equality* (Cambridge: Polity), p. 45.
31. For two arguments in favour of exemptions for minority groups, but against granting those groups complete license to act as they please, see J. Quong (2006) 'Cultural Exemptions, Expensive Tastes and Equal Opportunities', *Journal of Applied Philosophy*, 23, 53–71; and P. Bou-Habib (2006) 'A Theory of Religious Accomodation', *Journal of Applied Philosophy*, 23, 109–26.
32. P. Casal (2003) 'Is Multiculturalism Bad for Animals?', *Journal of Political Philosophy*, 11, 1–22, especially p. 17.
33. Ibid., p. 19.

6   Marxism and Animals

1. This account is largely drawn from K. Marx and F. Engels (1994) *The Communist Manifesto* and *The German Ideology Part 1* (selections) in L. Simon (ed.) *Karl Marx: Selected Writings* (Indianapolis, IN: Hackett).
2. This point is also made in Peter Singer (2000) *Marx: A Very Short Introduction* (Oxford: Oxford University Press), p. 81.
3. See, for example, K. Marx (1994) *Economic and Philosophic Manuscripts* (selections) in *Karl Marx: Selected Writings*, p. 63.
4. Ibid., pp. 58–66.
5. Ibid., p. 62.
6. Marx and Engels, *The German Ideology Part 1*, p. 152.
7. Marx, *Economic and Philosophic Manuscripts*, p. 64.
8. D. Sztybel (1997) 'Marxism and Animal Rights', *Ethics and the Environment*, 2, 169–85, especially, p. 175.
9. Ibid.
10. T. Benton (1993) *Natural Relations: Ecology, Animal Rights and Social Justice* (London: Verso), p. 40.

11. Ibid., p. 35.
12. Ibid., p. 36.
13. R. Garner (2005) *The Political Theory of Animal Rights* (Manchester: Manchester University Press), pp. 102–3.
14. H. Kean (1998) *Animal Rights: Political and Social Change in Britain Since 1800* (London: Reaktion Books), p. 124.
15. Garner, *The Political Theory of Animal Rights*, p. 101.
16. H. Parsons (1977) *Marx and Engels on Ecology* (Westport, CT: Greenwood Press), p. 47.
17. For one important work in this spirit, see K. Tester (1991) *Animals and Society: The Humanity of Animal Rights* (London: Routledge).
18. Parsons, *Marx and Engels on Ecology*, pp. 47–8.
19. M. Midgley (1983) *Animals and Why They Matter* (Athens, GA: University of Georgia Press), p. 23.
20. K. Perlo (2002) 'Marxism and the Underdog', *Society and Animals*, 10, 303–18, especially, p. 306.
21. Ibid., p. 306.
22. B. Noske (1997) *Beyond Boundaries: Humans and Animals* (Montreal, Quebec: Black Rose Books), p. 18.
23. Ibid., p. 19.
24. Ibid.
25. Ibid., p. 20.
26. M. Nussbaum (2006) *Frontiers of Justice: Disability, Nationality, Species Membership* (London: The Belknap Press of Harvard University Press), p. 345.
27. Garner, *The Political Theory of Animal Rights*, p. 109.
28. K. Marx (1994) 'Critique of the Gotha Program' in *Karl Marx: Selected Writings*, p. 321.
29. Sztybel, 'Marxism and Animal Rights', p. 165.
30. Benton, *Natural Relations*, p. 212.
31. For an interpretation along these lines, see Singer, *Marx: A Very Short Introduction*, p. 84.
32. Sztybel, 'Marxism and Animal Rights', p 179.
33. Ibid.
34. Benton, *Natural Relations*, p. 212.
35. Sztybel, 'Marxism and Animal Rights', p. 180.
36. Benton, *Natural Relations*, p. 165.
37. Ibid., p. 147.

7   Feminism and Animals

1. R. Garner (2005) *The Political Theory of Animal Rights* (Manchester: Manchester University Press), p. 141; J. Donovan and C. Adams (2007) 'Introduction' in J. Donovan and C. Adams (eds) *The Feminist Care Tradition in Animal Ethics* (New York: Columbia University Press), p. 8.
2. L. Gruen (2007) 'Empathy and Vegetarian Commitments' in *The Feminist Care Tradition in Animal Ethics*, p. 336.

3. J. Donovan (2007) 'Animal Rights and Feminist Theory' in *The Feminist Care Tradition in Animal Ethics*, p. 65.
4. See, for example, K. Warren (1990) 'The Power and the Promise of Ecological Feminism', *Environmental Ethics*, 12, 125–46; V. Plumwood (1991) 'Nature, Self, and Gender: Feminism, Environmental Philosophy, and the Critique of Rationalism', *Hypatia*, 6, 3–27.
5. C. Adams (2000) *The Sexual Politics of Meat: A Feminist-Vegetarian Critical Theory*, 10th anniversary edn (New York: Continuum), pp. 36–45.
6. C. MacKinnon (2007) 'Of Mice and Men: A Fragment on Animal Rights' in *The Feminist Care Tradition in Animal Ethics*, p. 319.
7. Adams, *The Sexual Politics of Meat*, p. 81.
8. Ibid., p. 66.
9. Ibid., p. 65.
10. MacKinnon, 'Of Mice and Men', p. 319.
11. See A. Cochrane (2009) 'Ownership and Justice for Animals', *Utilitas*, 21, 424–42.
12. P. Singer (1995) *Animal Liberation*, 2nd edn (London: Pimlico), p. xi.
13. T. Kelch (2007) 'The Role of the Rational and the Emotive in a Theory of Animal Rights' in *The Feminist Care Tradition in Animal Ethics*, p. 279.
14. Ibid., p. 280.
15. B. Luke (2007) 'Justice, Caring, and Animal Liberation' in *The Feminist Care Tradition in Animal Ethics*, p. 128.
16. M. Kheel (2007) 'The Liberation of Nature: A Circular Affair' in *The Feminist Care Tradition in Animal Ethics*, p. 47.
17. Luke, 'Justice, Caring, and Animal Liberation', p. 133.
18. Kheel, 'The Liberation of Nature', p. 51.
19. D. Slicer (2007) 'Your Daughter or Your Dog? A Feminist Assessment of the Animal Research Issue' in *The Feminist Care Tradition in Animal Ethics*, p. 108.
20. Ibid., pp. 108–9.
21. Ibid., p. 110.
22. Donovan, 'Animal Rights and Feminist Theory', p. 64.
23. C. Gilligan (1982) *In a Different Voice: Psychological Theory and Women's Development* (Cambridge, MA: Harvard University Press).
24. This account has been informed by R. Tong and N. Williams (2009) 'Feminist Ethics' in E.N. Zalta (ed.) *Stanford Encyclopedia of Philosophy*, available at http://plato.stanford.edu/archives/fall2009/entries/feminism-ethics/ (accessed 18 June 2010).
25. N. Noddings (1984) *Caring: A Feminine Approach to Ethics and Moral Education* (Berkeley, CA: University of California Press), p. 154.
26. Ibid., p. 156.
27. Josephine Donovan (2007) 'Attention to Suffering' in *The Feminist Care Tradition in Animal Ethics*, p. 185.
28. Ibid., p. 187.
29. Luke, 'Justice, Caring, and Animal Liberation', pp. 136–46.
30. Ibid., p. 135.
31. Kheel, 'The Liberation of Nature', p. 48.
32. T. Regan (1995) 'Obligations to Animals are Based on Rights', *Journal of Agricultural and Environmental Ethics*, 8, 171–80, especially, p. 177.

33. J. Franklin (2005) *Animal Rights and Moral Philosophy* (New York: Columbia University Press), p. 80.

8 Conclusions

1. J.S. Mill (1859) *On Liberty* in J. Gray (ed.) *John Stuart Mill: On Liberty and Other Essays* (Oxford: Oxford University Press, 1998), p. 14.
2. T. Machan (2002) 'Why Human Beings May Use Animals', *Journal of Value Inquiry*, 36, 9–14.

Bibliography

Aaltola, E. (2005) 'Animal Ethics and Interest Conflicts', *Ethics and the Environment*, 10, 21–48.

Abbey, R. (2007) 'Rawlsian Resources for Animal Ethics', *Ethics and the Environment*, 12, 1–22.

Adams, C. (2000) *The Sexual Politics of Meat: A Feminist-Vegetarian Critical Theory*, 10th anniversary edn (New York: Continuum).

Aquinas, T. (1928) *Summa Contra Gentiles*, trans. The English Dominican Fathers (London: Burns Oates and Washbourne).

Aquinas, T. (1929) *Summa Theologica*, trans. Fathers of the English Dominican Province (London: Burns Oates and Washbourne).

Aristotle (1991) *The Nicomachean Ethics*, trans. D. Ross (Oxford: Oxford University Press).

Aristotle (1992) *The Politics*, trans. T. Sinclair, revised edn (London: Penguin).

Augustine (1998) *The City of God against the Pagans*, trans. R. Dyson (Cambridge: Cambridge University Press).

Barry, B. (1989) *Theories of Justice* (Hemel Hempstead: Harvester-Wheatsheaf).

Barry, B. (1995) *Justice as Impartiality* (Oxford: Oxford University Press).

Barry, B. (2001) *Culture and Equality* (Cambridge: Polity).

Bekoff, M. and J. Pierce (2009) *Wild Justice: The Moral Lives of Animals* (Chicago, IL: University of Chicago Press).

Bentham, J. (1780) *An Introduction to the Principles of Morals and Legislation* (Oxford: Clarendon Press, 1823).

Benton, T. (1993) *Natural Relations: Ecology, Animal Rights and Social Justice* (London: Verso).

Bittman, M. (2008) 'Rethinking the Meat-guzzler', *New York Times*, 27 January.

Bou-Habib, P. (2006) 'A Theory of Religious Accommodation', *Journal of Applied Philosophy*, 23, 109–26.

Callicott, J. (1989) 'Animal Liberation and Environmental Ethics: Back Together Again' in J. Baird Callicott, *In Defense of the Land Ethic: Essays in Environmental Philosophy* (Albany, NY: State University of New York Press), pp. 49–62.

Callicott, J. (1998) ' "Back Together Again" Again', *Environmental Values*, 7, 461–75.

Carone, G. (2001) 'The Classical Greek Tradition' in D. Jamieson (ed.) *A Companion to Environmental Philosophy* (Oxford: Blackwell), pp. 67–80.

Carruthers, P. (1992) *The Animals Issue: Moral Theory in Practice* (Cambridge: Cambridge University Press).

Casal, P. (2003) 'Is Multiculturalism Bad for Animals?', *Journal of Political Philosophy*, 11, 1–22.

Cataldi, S. (2002) 'Animals and the Concept of Dignity: Critical Reflections on a Circus Performance', *Ethics and the Environment*, 7, 104–26.

Cavalieri, P. and P. Singer (1993) (eds) *The Great Ape Project: Equality Beyond Humanity* (London: Fourth Estate).

Cochrane, A. (2007) 'Animal Rights and Animal Experiments: An Interest-Based Approach', *Res Publica*, 13, 293–318.

Cochrane, A. (2009) 'Do Animals Have an Interest in Liberty?', *Political Studies*, 57, 660–79.

Cochrane, A. (2009) 'Ownership and Justice for Animals', *Utilitas*, 21, 424–42.

Cohen, C. (1986) 'The Case for the Use of Animals in Biomedical Research', *New England Journal of Medicine*, 315, 865–70.

Davis, S. (2003) 'The Least Harm Principle May Require that Humans Consume a Diet Containing Large Herbivores, not a Vegan Diet', *Journal of Agricultural and Environmental Ethics*, 16, 387–97.

DEFRA (2007) 'Survey of Public Attitudes and Behaviours toward the Environment: 2007', http://www.defra.gov.uk/evidence/statistics/environment/pubatt/index.htm.

Devine, P. (1978) 'The Moral Basis of Vegetarianism', *Philosophy*, 53, 481–505.

Dombrowski, D. (1984) *The Philosophy of Vegetarianism* (Amherst, MA: University of Massachusetts Press).

Donovan, J. (2007) 'Animal Rights and Feminist Theory' in J. Donovan and C. Adams (eds) *The Feminist Care Tradition in Animal Ethics* (New York: Columbia University Press), pp. 58–86.

Donovan, J. (2007) 'Attention to Suffering' in J. Donovan and C. Adams (eds) *The Feminist Care Tradition in Animal Ethics* (New York: Columbia University Press), pp. 174–197.

Donovan, J. and C. Adams (2007) (eds) *The Feminist Care Tradition in Animal Ethics* (New York: Columbia University Press).

Dworkin, R. (1973) 'The Original Position', *University of Chicago Law Review*, 40, 500–33.

Elliot, R. (1984) 'Rawlsian Justice and Non-human Animals', *Journal of Applied Philosophy*, 1, 95–106.

Feinberg, J. (1974) 'The Rights of Animals and Unborn Generations' in W. Blackstone (ed.) *Philosophy and Environmental Crisis* (Athens, GA: University of Georgia Press), pp. 43–68.

Fox, M. (1978) 'Animal Liberation: A Critique', *Ethics*, 88, 106–18.

Francione, G. (1996) *Rain Without Thunder: The Ideology of the Animal Rights Movement* (Philadelphia, PA: Temple University Press).

Francione, G. (2004) 'Animals: Property or Persons?' in C. Sunstein and M. Nussbaum (eds) *Animal Rights: Current Debates and New Directions* (Oxford: Oxford University Press), pp. 108–142.

Francione, G. (2008) *Animals as Persons: Essays on the Abolition of Animal Exploitation* (New York: Columbia University Press).

Franklin, J. (2005) *Animal Rights and Moral Philosophy* (New York: Columbia University Press).

Frey, R. (1980) *Interests and Rights* (Oxford: Clarendon Press).

Garner, R. (1993) *Animals, Politics and Morality* (Manchester: Manchester University Press).

Garner, R. (2003) 'Animals, Politics and Justice: Rawlsian Liberalism and the Plight of Non-humans', *Environmental Politics*, 12, 3–22.

Garner, R. (2005) *The Political Theory of Animal Rights* (Manchester: Manchester University Press).

Garrett, A. (2007) 'Frances Hutcheson and the Origin of Animal Rights', *Journal of the History of Philosophy*, 45, 243–65.

Gilligan, C. (1982) *In a Different Voice: Psychological Theory and Women's Development* (Cambridge, MA: Harvard University Press).

Glendinning, L. (2008) 'Spanish Parliament Approves "Human Rights" for Apes', *Guardian*, 26 June.

Griffin, J. (2008) *On Human Rights* (Oxford: Oxford University Press).

Gruen, L. (2007) 'Empathy and Vegetarian Commitments' in J. Donovan and C. Adams (eds) *The Feminist Care Tradition in Animal Ethics* (New York: Columbia University Press), pp. 333–343.

Hadley, J. (2007) 'Critique of Callicott's Biosocial Moral Theory', *Ethics and the Environment*, 12, 67–78.

Harrison, R. (1964) *Animal Machines: The New Factory Farming Industry* (London: V. Stuart).

Hobbes, T. (1651) *Leviathan*, ed. C. Macpherson (London: Penguin, 1968).

Hobbes, T. (1651) *De Cive* in B. Gent (ed.) *Man and Citizen* (Indiana, IN: Hackett Publishing, 1991).

IPSOS-MORI (2009) 'Views on Animal Experimentation', http://www.ipsos-mori.com/Assets/Docs/Polls/views-on-animal-experimentation-2009.pdf.

Kant, I. (1963) *Lectures on Ethics*, trans. L Infield (New York: Harper and Row).

Kean, H. (1998) *Animal Rights: Political and Social Change in Britain Since 1800* (London: Reaktion Books).

Kelch, T. (2007) 'The Role of the Rational and the Emotive in a Theory of Animal Rights' in J. Donovan and C. Adams (eds) *The Feminist Care Tradition in Animal Ethics* (New York: Columbia University Press), pp. 259–300.

Kheel, M. (2007) 'The Liberation of Nature: A Circular Affair' in J. Donovan and C. Adams (eds) *The Feminist Care Tradition in Animal Ethics* (New York: Columbia University Press), pp. 39–57.

Kukathas, C. (1997) 'Cultural Toleration' in I. Shapiro and W. Kymlicka (eds) *NOMOS XXIX: Ethnicity and Group Rights* (New York: New York University Press), pp. 69–104.

Linzey, A. (1994) *Animal Theology* (London: SCM Press).

Lo, Y. (2001) 'The Land Ethic and Callicott's Ethical System (1980–2001): An Overview and Critique', *Inquiry*, 44, 331–58.

Locke, J. (1690) *Two Treatises on Government*, ed. P. Laslett, student edn (Cambridge: Cambridge University Press, 1988).

Luke, B. (2007) 'Justice, Caring, and Animal Liberation' in J. Donovan and C. Adams (eds) *The Feminist Care Tradition in Animal Ethics* (New York: Columbia University Press), pp. 125–152.

Machan, T. (2002) 'Why Human Beings May Use Animals', *Journal of Value Inquiry*, 36, 9–14.

Machiavelli, N. (1532) *The Prince*, eds Q. Skinner and R. Price (Cambridge: Cambridge University Press, 1988).

MacIntyre, A. (1981) *After Virtue: A Study in Moral Theory* (London: Duckworth).

MacKinnon, C. (2007) 'Of Mice and Men: A Fragment on Animal Rights' in J. Donovan and C. Adams (eds) *The Feminist Care Tradition in Animal Ethics* (New York: Columbia University Press), pp. 316–332.

Marx, K. (1994) *Economic and Philosophic Manuscripts* (selections) in L. Simon (ed.) *Karl Marx: Selected Writings* (Indianapolis, IN: Hackett).

Marx, K. (1994) 'Critique of the Gotha Program' in L. Simon (ed.) *Karl Marx: Selected Writings* (Indianapolis, IN: Hackett).

Marx, K. and F. Engels (1994) *The Communist Manifesto* in L. Simon (ed.) *Karl Marx: Selected Writings* (Indianapolis, IN: Hackett).

Marx, K. and F. Engels (1994) *The German Ideology Part 1* (selections) in L. Simon (ed.) *Karl Marx: Selected Writings* (Indianapolis, IN: Hackett).

Matheny, G. (2003) 'Least Harm: A Defense of Vegetarianism from Steven Davis's Omnivorous Proposal', *Journal of Agricultural and Environmental Ethics*, 16, 505–11.

Matheny, G. (2005) 'Utilitarianism and Animals' in P. Singer (ed.) *In Defense of Animals: The Second Wave* (Oxford: Blackwell), pp. 505–511.

McCloskey, H. (1965) 'Rights', *The Philosophical Quarterly*, 15, 115–27.

Midgley, M. (1983) *Animals and Why They Matter* (Athens, GA: University of Georgia Press).

Mill, J.S. (1859) *On Liberty* in J. Gray (ed.) *John Stuart Mill: On Liberty and Other Essays* (Oxford: Oxford University Press, 1998).

Moller Okin, S., J. Cohen, M. Howard, M. Nussbaum (1999) *Is Multiculturalism Bad for Women?* (Princeton, NJ: Princeton University Press).

Nobis, N. (2004) 'Carl Cohen's "Kind" Arguments For Animal Rights and Against Human Rights', *Journal of Applied Philosophy*, 21, 43–59.

Noddings, N. (1984) *Caring: A Feminine Approach to Ethics and Moral Education* (Berkeley, CA: University of California Press).

Noske, B. (1997) *Beyond Boundaries: Humans and Animals* (Montreal, Quebec: Black Rose Books).

Nussbaum, M. (2006) *Frontiers of Justice: Disability, Nationality, Species Membership* (London: The Belknap Press of Harvard University Press).

Ovid (2004) *Metamorphoses*, trans. D. Raeburn (London: Penguin).

Parekh, B. (2000) *Rethinking Multiculturalism: Cultural Diversity and Political Theory* (Houndmills, Basingstoke: Palgrave).

Parsons, H. (1977) *Marx and Engels on Ecology* (Westport, CT: Greenwood Press).

Passmore, J. (1975) 'The Treatment of Animals', *Journal of the History of Ideas*, 36, 195–218.

Perlo, K. (2002) 'Marxism and the Underdog', *Society and Animals*, 10, 303–18.

Petrinovich, L. (1999) *Darwinian Dominion: Animal Welfare and Human Interests* (London: MIT Press).

Plato (2003) *The Republic*, trans. D. Lee, 2nd edn (London: Penguin).

Pluhar, E. (1995) *The Moral Significance of Human and Nonhuman Animals* (Durham, NJ: Duke University Press).

Plumwood, V. (1991) 'Nature, Self, and Gender: Feminism, Environmental Philosophy, and the Critique of Rationalism', *Hypatia*, 6, 3–27.

Porphyry (1965) *On Abstinence from Animal Food*, trans. T. Taylor (London: Centaur Press).

Preece, R. (2008) *Sins of the Flesh: A History of Ethical Vegetarian Thought* (Vancouver: UBC Press).

Preece, R. and L. Chamberlain (1993) *Animal Welfare and Human Values* (Waterloo, Ontario: Wilfrid Laurier University Press).

Primatt, H. (1992) *The Duty of Mercy and the Sin of Cruelty to Brute Animals*, eds R. Ryder and J. Baker (Fontwell, Sussex: Centaur Press).

Pritchard, M. and W. Robison (1981) 'Justice and the Treatment of Animals: A Critique of Rawls', *Environmental Ethics*, 3, 55–61.

Quong, J. (2006) 'Cultural Exemptions, Expensive Tastes and Equal Opportunities', *Journal of Applied Philosophy*, 23, 53–71.

Rawls, J. (1999) *A Theory of Justice*, revised edn (Oxford: Oxford University Press).

Regan, T. (1980) 'Utilitarianism, Vegetarianism and Animal Rights', *Philosophy and Public Affairs*, 9, 305–37.

Regan, T. (1995) 'Obligations to Animals Are Based on Rights', *Journal of Agricultural and Environmental Ethics*, 8, 171–80.

Regan, T. (2004) *The Case for Animal Rights*, 2nd edn (Berkeley, CA: Berkeley University Press).

Reinders, E. (2000) 'Animals, Attitude Toward: Buddhist Perspectives' in W. Johnson (ed.) *Encyclopedia of Monasticism Vol. 1* (Chicago, IL: Fitzroy Dearborn Publishers), pp. 30–31.

Rousseau, J-J. (1755) *A Discourse on Inequality*, trans. M. Cranston (London: Penguin, 1984).

Rousseau, J-J. (1762) *The Social Contract*, trans. M. Cranston (London: Penguin, 1968).

Rowlands, M. (1997) 'Contractarianism and Animal Rights', *Journal of Applied Philosophy*, 14, 235–47.

Russell, B. (1994) *History of Western Philosophy: and its Connection with Political and Social Circumstances from the Earliest Times to the Present Day*, 2nd edn (London: Routledge).

Sandel, M. (1982) *Liberalism and the Limits of Justice* (Cambridge: Cambridge University Press).

Schedler, G. (2005) 'Does Ethical Meat Eating Maximize Utility?', *Social Theory and Practice*, 31, 499–511.

Serpell, J. and E. Paul (1994) 'Pets and the Development of Positive Attitudes to Animals' in A. Manning and J. Serpell (eds) *Animals and Human Society: Changing Perspectives* (London: Routledge), pp. 127–144.

Singer, B. (1988) 'An Extension of Rawls' Theory of Justice to Environmental Ethics', *Environmental Ethics*, 10, 217–31.

Singer, P. (1972) 'Famine, Affluence, and Morality', *Philosophy and Public Affairs*, 1, 229–43.

Singer, P. (1987) 'Animal Liberation or Animal Rights?', *The Monist*, 70, 3–14.

Singer, P. (1993) *Practical Ethics*, 2nd edn (Cambridge: Cambridge University Press).

Singer, P. (1995) *Animal Liberation*, 2nd edn (London: Pimlico).

Singer, P. (2000) *Marx: A Very Short Introduction* (Oxford: Oxford University Press).

Singer, P. (2002) 'A Response to Martha Nussbaum', http://www.utilitarian.net.

Slicer, D. (2007) 'Your Daughter or Your Dog? A Feminist Assessment of the Animal Research Issue' in J. Donovan and C. Adams (eds) *The Feminist Care Tradition in Animal Ethics* (New York: Columbia University Press), pp. 105–124.

Sorabji, R. (1993) *Animal Minds and Human Morals: The Origins of the Western Debate* (New York: Cornell University Press).

Steiner, G. (2005) *Anthropocentrism and its Discontents: The Moral Status of Animals in the History of Western Philosophy* (Pittsburgh, PA: University of Pittsburgh Press).

Sztybel, D. (1997) 'Marxism and Animal Rights', *Ethics and the* Environment, 2, 169–85.

Taylor, C. (1985) *Philosophy and the Human Sciences* (Cambridge: Cambridge University Press).

Taylor, P. (1986) *Respect for Nature: A Theory of Environmental Ethics* (Princeton, N J: Princeton University Press).

Tester, K. (1991) *Animals and Society: The Humanity of Animal Rights* (London: Routledge).

Tong, R. and N. Williams (2009) 'Feminist Ethics' in E.N. Zalta (ed.) *Stanford Encyclopedia of Philosophy*, http://plato.stanford.edu/archives/fall2009/entries/feminism-ethics/.

VanDeVeer, D. (1979) 'Of Beasts, Persons, and the Original Position', *The Monist*, 62, 368–77.

Waldron, J. (2005) 'Torture and Positive Law: Jurisprudence for the White House', *Columbia Law Review*, 105, 1681–750.

Walzer, M. (1983) *Spheres of Justice: A Defense of Pluralism and Equality* (New York: Basic Books).

Warren, K. (1990) 'The Power and the Promise of Ecological Feminism', *Environmental Ethics*, 12, 125–46.

Watts, J. (2009) 'China Plans First Animal Welfare Law', *Guardian*, 16 June.

White, L. (1967) 'The Historical Roots of Our Ecological Crisis', *Science*, 155, 1203–7.

Williams, J. (2002) 'US Judge Blocks Tribal Whaling', *BBC News Online*, 4 May, http://news.bbc.co.uk/1/hi/world/americas/1967677.stm.

Wise, S. (2002) *Unlocking the Cage: Science and the Case for Animal Rights* (Oxford: Perseus Press).

Index

Note: Locators followed by 'n' refers to note numbers cited in the text.

Aaltola, Elisa, 82, 152 n. 18
Abbey, Ruth, 61, 150 n. 16
Adams, Carol, 119–20, 154 n. 1, 155
    n. 5, nn. 7–9
agriculture, 4, 31–42, 105–7
    crop-only system, 41–2
    crops and ruminant pasture system,
        41–2
    intensive agriculture, 105–7
    see also farming; meat, the meat
        industry
animal experimentation, 34–5, 45, 48,
    80, 129, 133, 143
    levels of support for, 80
    non-therapeutic, 34
    therapeutic, 34
animal sacrifice, 61
Animal Welfare Act (2007), 1,
    2, 75
anthropocentrism, 99–101
Aquinas, Thomas, 4–5, 16–20, 23, 27,
    116, 148 nn. 21–2, n. 28
Aristotle, 4–6, 13–16, 17–20, 27, 73,
    116, 118, 147 nn. 10–12
    moral and intellectual virtues, 13
    telos, 13
Assisi, St. Francis of, 19, 78
Augustine, 16–20, 27, 116, 148 n. 20,
    n. 27
Aurelius, Marcus, 14
autonomy, 23–5, 51, 59–61, 65–8,
    75, 125
    Kantian autonomy, 66–7
    preference autonomy, 66–7
    see also conception of the good;
        dignity; ends-in-themselves;
        personhood

badger-baiting, 79–80, 88
Barry, Brian, 151 n. 20, 152 n. 20,
    153 n. 30
bear-baiting, 79, 88
bear dancing, 43, 45, 48
bear farming, 78
Bekoff, Marc, 150 n. 13
Bentham, Jeremy, 6, 16, 24–7, 28–32,
    39, 50, 51, 96, 148 n. 42
Benton, Ted, 109–13, 153–4
    nn. 10–12, 154 n. 30, n. 34,
    nn. 36–7
Bou-Habib, Paul, 153 n. 31

Callicott, James Baird, 83–5, 152–3
    nn. 21–2, 153 n. 24
    biotic community, 83
capitalism, 93–4, 97
    effects on animals, 105–8
    see also Marx, Karl
Carone, Gabriela, 147 n. 9
Carruthers, Peter, 58–9, 150 nn. 11–12
Casal, Paula, 90–1, 153 nn. 32–3
Cataldi, Suzanne, 149 n. 19
Cavalieri, Paula, 151 n. 25
charity, to animals, 17, 19
    as distinct from justice, 3
China, animal welfare legislation, 1–4
China, attitudes to animal welfare,
    78–9
Chryssipus, 14
Cicero, 4–5, 14, 116
circuses, 43
Cobbe, Frances Power, 115
Cochrane, Alasdair, 150 n. 34, 151
    n. 32, nn. 34–5, 155 n. 11
cock-fighting, 79–80

Cohen, Carl, 45–6, 150 n. 27
common good, 72–4, 78, 85, 91–2, 95
communism, *see* Marx, Karl
communitarianism, 4, 7, 27, 72–92,
    93, 136–7, 139–40, 144
conception of the good, 7, 52, 54–5,
    61, 70
    see also autonomy; dignity;
        ends-in-themselves;
        personhood
consciousness, 36, 40, 42–3, 63, 85–9,
    111, 125, 128, 133, 138
    'merely conscious' entities, 40–1, 45
    self-conscious entities, 40–1, 45
    see also Singer, Peter
consequentialism, 38–42
    see also utilitarianism
contract, 7, 22–4, 27, 29, 52–65, 68,
    83–4

Darwin, Charles, 99, 146
Davis, Steven, 41, 149 n. 16
Descartes, 21, 24, 26–7
Devine, Philip, 149 n. 12
dignity, 23–4, 51, 60
    see also autonomy; conception of
        the good; ends-in-themselves;
        personhood
Dombrowski, Dan, 147 n. 7
dominion, man's over Earth,
    17–20, 27
Donovan, Josephine, 118, 132–3, 154
    n. 1, 155 n. 3, n. 22, n. 27
Dworkin, Ronald, 68–70, 151 n. 36
    equal concern and respect, 68–9

ecofeminism, 118
egalitarianism, 30–1, 41, 51, 138
    see also utilitarianism; liberalism
Elliot, Robert, 150 n. 11, 151 n. 18,
    n. 21
Empedocles, 15
ends-in-themselves, 23, 51, 66–7
    see also autonomy; conception of
        the good; ends-in-themselves;
        personhood
Engels, Friedrich, 104, 153 n. 1, n. 6
experimentation, *see* animal
    experimentation

Farm Animal Welfare Council
    (FAWC), 86
farming, 3, 5, 31–42, 141, 143
    factory, 31–4, 39, 48, 76, 83–4, 90–1,
        107, 129, 143
    free-range, 39–42, 75, 84
    intensive, 75–80, 90, 105–7, 111
    see also agriculture; meat, the meat
        industry
Feinberg, Joel, 47–8, 150 n. 33
feminism, 4, 8, 28, 115–17
    care-based, 8–9, 28, 115–35, 137,
        141, 144
    liberal feminism, 116
    radical feminism, 116
foie gras, 75, 82
Fox, Michael, 45–6, 150 n. 27
Francione, Gary, 66–8, 151 nn. 29–31
Franklin, Julian, 156 n. 33
Frey, R. G., 36–7, 149 n. 9

Gandhi, 105
Garner, Robert, 63–5, 102, 149 n. 5,
    150 n. 15, 151 nn. 23–4, n. 33,
    n. 35, 154 n. 13, n. 15, n. 27, n. 1
Garrett, Aaron, 148 n. 40
Gilligan, Carol, 130–1, 155 n. 23
great apes, 1–2, 59, 65
Griffin, James, 148 n. 39
Gruen, Lori, 154 n. 2

Hadley, John, 54–5, 153 n. 26
Harrison, Ruth, 32, 149 n. 1
Hobbes, Thomas, 22, 52, 148
    nn. 33–4
honour killing, 87
Humanitarian League, 102
hunting, 70, 78, 119
    aboriginal, 85–91
    with hounds, 80
Hunting Act (2004), 80
Hutcheson, Frances, 6, 24, 26, 29, 50

impartiality, 54, 62, 64, 81–5, 128,
    130–3
interests, 7, 9, 24, 30–40, 47–8, 49, 63,
    68–9, 81–4, 87–9, 101–5, 112,
    120, 128–30, 138, 143–5
    interest-based rights, *see* rights

whether possessed by animals, 35–7
see also utility, definitions of; welfare
International Whaling Commission
(IWC), 86

justice, 1–9, 11–28, 29–32, 44, 52–65,
69–70, 74–6, 77, 95, 108, 109–13,
115–17, 123–34, 136–41, 142,
144, 145–6
for animals, definition, 3–4
justice as fairness, *see* Rawls,
John

Kant, Immanuel, 23–4, 26, 45, 51, 60,
65–8, 116, 148 n. 38
Kean, Hilda, 154 n. 14
Kelch, Thomas, 124, 155 nn. 13–14
Kheel, Marti, 127, 155 n. 16, n. 18,
n. 31
Kingsford, Anna, 115
Kohlberg, Lawrence, 130–1
Kukathas, Chandran, 87–8, 153 n. 28

language, 118–22
as a tool to degrade women and
animals, 119–20, 122
liberalism, 4, 7, 9, 27, 49, 50–71, 72–3,
75, 93, 136, 139, 141–5
Linzey, Andrew, 148 n. 26
Lo, Norva, 153 n. 25
Locke, John, 22, 52, 116,
148 n. 35
Luke, Brian, 126, 133–4, 155 n. 15,
n. 17, n. 29

MacDonald, Ramsey, 102
Machan, Tibor, 156 n. 2
Machiavelli, Niccolò, 20, 148 n. 31
MacIntyre, Alasdair, 72, 152 n. 1
MacKinnon, Catherine, 119, 155
n. 16, n. 10
marginal cases, argument from, 125
Martin's Act (1822), 75
Marx, Karl, 4–5, 8, 25–6, 93–115, 118,
137, 140–1, 144, 153 n.1, nn. 3–7,
154 n. 28
alienation, 96–7: of animals, 105–8
communism, 26, 94–7, 100,
110, 114

exploitation, 117–118: of animals,
105–8
historical materialism, 94–5,
96–101
modes of production, 94, 114
productive forces, 26, 94, 109
relations of production, 94–5
species-being, 96–8, 106
superstructure, 94–5
Matheny, Gaverick, 149 n. 13, n. 17
McCloskey, H.J., 36, 149 n. 8
meat, 2, 11, 13, 31–42, 70, 85–9, 107,
119–22, 133, 137
halal, 85–6, 89
kosher, 85–6, 89
meat consumption as a masculine
trait, 119, 121–2
meat eating as a human interest,
33–4, 38–9
rates of meat consumption, 2, 39
the meat industry, 3, 31–42, 45,
48–9, 122
see also agriculture; farming
Midgley, Mary, 81–3, 152 nn. 12–17,
154 n. 19
mixed community, 82–4
Mill, John Stuart, 6, 25–6, 29–30, 69,
105, 143, 151 n. 37, 156 n. 1
harm principle, 69, 143
Moller Okin, Susan, 152 n. 5
multiculturalism, 85–91

natural law, 4, 14, 18
Nazis, 103–4
attitude to animal welfare, 103–4
needs, 8, 63, 68–9, 95, 97, 108–12,
128, 130
as a condition for meriting justice,
8, 108–12
see also Marx, Karl
Nobis, Nathan, 46, 150 n. 29
Noddings, Nel, 131–2, 155 nn. 25–6
non-native species, 83, 100
Noske, Barbara, 106–8, 154
nn. 22–5
Nussbaum, Martha, 42–4, 106, 149
n. 18, n. 20, nn. 22–3, 154 n. 26
capabilities, 43–4, 106
functionings, 43–4, 106, 108

objectification, 118, 120, 122–3
oppression, 102, 116, 117–23
  linked oppressions of women and
    animals, 117–23
Orwell, George, 76, 80
Ovid, 11, 147 n. 1

Parekh, Bhikhu, 77–8, 152 n. 6
Parsons, Howard, 102–3, 154 n. 16,
  n. 18
partiality, 81–5, 130–33
  first-order, 82–3
  second-order, 82–3
particularism, 76–9
Passmore, John, 148 n. 23
Paul, Elizabeth, 152 n. 3, n. 8
payback system in law, 87
Perlo, Katherine, 105–8, 154
  nn. 20–1
personhood, 51–62, 65–70, 138
  as a condition for meriting justice,
    57–60
  whether possessed by animals, 59,
    65–8
  see also autonomy; conception of
    the good; ends-in-themselves;
    dignity
Petrinovich, Lewis, 37, 149
  nn. 10–11
philosopher-kings, 12
Pierce, Jessica, 150 n. 13
Plato, 12–13, 14, 18, 116, 118, 147
  nn. 5–6, n. 8
Pluhar, Evelyn, 66, 151 n. 28
Plumwood, Val, 155 n. 4
pluralism, 7, 52, 60–1
Plutarch, 6, 15–16, 27, 96
Porphyry, 6, 15–16, 24, 27, 96, 148
  n. 16, n. 19
Preece, Rod, 147 n. 3, 148 n. 24
Primatt, Humphrey, 6, 24, 29, 50, 148
  n. 41
Pritchard, Michael, 151 n. 17
property status, 66–8, 118, 120,
  122–3
Pythagoras, 10–12, 15
  Pythagorean diet, 11

Quong, Jonathan, 153 n. 31

Rawls, John, 50–70, 72, 75, 110, 116,
  150 nn. 1–10, n. 14, 151 n. 22
  difference principle, 55, 60
  justice as fairness, 55, 68
  liberty principle, 55, 60–1
  original position, 54, 60–5
  reflective equilibrium, 63–5
  veil of ignorance, 54–5, 63–5
reason, 13–23, 27–9, 116–18, 121–30
  as a condition for meriting justice,
    13–20, 27, 29, 116
  as a means of oppression, 118, 121
  as a method in political theory, 116,
    123–30
  whether possessed by animals,
    15–16, 21, 23
reciprocity, 46–7, 56
Regan, Tom, 45–50, 66, 79, 126, 149
  n. 12, 150 n. 25, nn. 30–2, 151
  nn. 18–19, n. 26, 155 n. 32
  inherent value, 45, 47, 50, 66, 79
  subjects-of-a-life, 45–7, 66, 126
reincarnation, 11, 15
  see also Pythagoras
Reinders, Eric, 152 n. 7
religious slaughter of animals,
  85–6, 90
respect, 11, 45–6, 60–1, 66–9, 77, 86–7
  see also Dworkin, Ronald
rights, 1–2, 7, 9, 14, 22, 24, 31, 44–8,
  61, 66–7, 73–6, 89, 101–5, 109,
  112–13, 116, 126–7, 140–5
  animal rights, 7, 9, 31, 44–8, 61,
    66–7, 101–5, 109, 112–13, 127,
    140–5
  critiques of rights, 109, 112–13, 127
  human rights, 1–2, 14, 24
  interest-based rights, 9, 47–8, 141–5
  natural rights, 22
  rights inflation, 143–4
Robison, Wade, 151 n. 17
Rousseau, Jean-Jacques, 22–3, 52, 148
  nn. 36–7
  general will, 22
Rowlands, Mark, 61–5, 151 nn. 18–19
Royal Society for the Prevention of
  Cruelty to Animals (RSPCA),
  1, 102
Russell, Bertrand, 12, 147 n. 4

Salt, Henry, 102, 105, 115
Sandel, Michael, 72, 151–2 n. 1
Schedler, George, 41, 149 n. 16
Seneca, 14
sentience, 4, 16, 21, 24–5, 29–32,
    43–4, 47, 50, 65, 112, 125, 138–9,
    142
  as a condition for meriting justice,
    4, 16, 24–5, 29–32, 47, 50, 65,
    125, 138–9, 142
  whether possessed by animals, 21
sentiment, 83, 116, 123–6, 130–4,
    141, 145
separateness of persons, 51, 70, 72–3
Serpell, James, 152 n. 3, n. 8
Shaftesbury, Lord, 102
Shaw, George Bernard, 102, 115
Singer, Brent, 151 n. 18
Singer, Peter, 5–7, 25, 30–45, 50, 81,
    84, 96, 123–30, 137–8, 141, 149
    nn. 2–4, nn. 6–7, nn. 14–15,
    n. 24, 151 n. 25, 153 n. 23, 153
    n. 2, 154 n. 31, 155 n. 12
  equal consideration of interests,
    31–5, 81, 84
  replacement argument, 39–41, 45
  speciesism, 31–5
Slicer, Deborah, 155 n. 19
social contract, *see* contract
Sorabji, Richard, 147 n. 2, nn. 13–14
sow stalls, 75
speciesism, *see* Singer, Peter
state of nature, 22, 52–4
Steiner, Gary, 147 n. 15, 148 n. 17,
    n. 29–30, n. 32
Stoics, 6, 13–20, 27, 116
Sztybel, David, 109–12, 153 nn. 8–9,
    154 n. 29, nn. 32–3, n. 35, n. 37

Taylor, Charles, 72, 152 n. 1
Taylor, Paul, 149 n. 21
Tester, Keith, 154 n. 17
Theophrastus, 6, 15, 96
Tong, Rosemarie, 155 n. 24
torture, 1–2, 45, 48
transmigration, *see* reincarnation
Treaty of Amsterdam, EU (1997), 3–4

UK, 1–2, 74–6, 79–80, 85–6, 89, 102
  attitudes to animal welfare, 74–6,
    79–80
universalism, 14, 22, 24, 73, 77–9,
    87–90, 98, 130
utilitarianism, 4, 6–9, 16, 24–5, 27,
    29–49, 50–1, 72–3, 138–9, 141–2
utility, definitions of, 29–31, 39
  see also interests; welfare

VanDeVeer, Donald, 62–5, 151
    nn. 18–19
veal crates, 75
veganism, rates of, 80
vegetarianism, 3, 11, 13, 16, 38–41,
    78, 80, 82, 89, 102, 121, 131, 137
  rates of, 80
Victorian social reform movement,
    101–3, 115
Victoria Street Society for the
    Protection of Animals Liable to
    Vivisection, 115
virtue, 4, 14, 73
  see also Aristotle
vivisection, 21, 115
  see also animal experimentation

Waldron, Jeremy, 150 n. 35
Walzer, Michael, 72, 76–7, 151 n. 1,
    152 n. 4
Warren, Karen, 155 n. 4
welfare, 6–7, 24, 30–1, 44, 45, 51, 53,
    63, 68–70, 72–3, 136, 138, 139,
    140, 142
  see also interests; utility,
    definitions of
welfarism, 30–1, 138, 142–3
  see also utilitarianism
whaling, 78, 86–90
White, Lynn, 148 n. 25
Williams, Nancy, 155 n. 24
Wise, Steven, 66, 151 n. 28

Xuan, Dao, 78

Yun, Xu, 78

Zeno, 14

BC 2/11